Introduction to Glass Integrated Optics

The Artech House Optoelectronics Library

Brian Culshaw, Alan Rogers, and Henry Taylor, *Series Editors*

Introduction to Glass Integrated Optics

S. Iraj Najafi
Editor

Artech House
Boston • London

Library of Congress Cataloging-in-Publication Data

Najafi, S. Iraj.
  Introduction to glass integrated optics / S. Iraj Najafi.
    p.   cm.
  Includes bibliographical references and index.
  ISBN 0-89006-547-0
  1. Integrated optics. 2. Optical wave guides. 3. Glass fibers.
  I. Title.
  TA1660.N35 1992   92-7350
  621.36'93--dc20   CIP

British Library Cataloguing in Publication Data

Najafi, S. Iraj
  Introduction to Glass Integrated Optics
  I. Title
  621.382

  ISBN 0-89006-547-0

© 1992 ARTECH HOUSE, INC.
685 Canton Street
Norwood, MA 02062

International Standard Book Number: 0-89006-547-0
Library of Congress Catalog Card Number: 92-7350

10 9 8 7 6 5 4 3 2 1

Contents

Chapter 1
Introduction
S. Iraj Najafi

Photonic Group of Montreal, École Polytechnique

1.1 INTEGRATED OPTICS

With the advent of the gas laser in 1960, its potential utilization as a coherent source for optical communication has been the subject of a broad range of research activities. Line-of-sight systems were demonstrated [1], but their usefulness was limited very much by weather conditions. Furthermore, gas lasers have serious limitations in their use for optical communication due to their size and need for high power. Optical communication became practical after two breakthroughs: an AlGaAs heterojunction was realized in 1969 and low-threshold current semiconductor lasers achieved shortly afterwards; optical fibers were produced [2].

To effectively use the semiconductor lasers and optical fibers, the need for components such as transmitters, receivers, modulators, and so forth soon became obvious. In 1969 S.E. Miller proposed the concept of "integrated optics" [3]. Integrated optics devices are now considered for applications in addition to optical communication (e.g., sensors, optical computers). A waveguide forms the basis for integrated optical devices. Different technologies have been employed to achieve waveguides in various materials. The devices made in these materials may be divided into four major groups: passive, electrooptic, optoelectronic, and all optical.

Passive integrated optical devices primarily split signals into two or more parts and route them in desired directions. Glass has been the most popular material to make passive components. Ion-exchange technique has been used extensively for waveguide fabrication in glass.

Integrated electrooptical devices are made in dielectric materials with large electrooptic coefficients. An electric field is used to modify the refractive index of the waveguide to control its operation. The most extensively developed technology for electrooptic devices uses $LiNbO_3$ substrates. The waveguides are formed by Ti:diffusion or by P-exchange.

Optoelectronic integrated optical devices are made of semiconducting material and offer the greatest versatility of any integrated optical component. The most ambitious goal of research on optoelectronic integrated devices is to develop a practical technology that can integrate optical and electrical circuits on the same chip. A lot of work has gone into making devices with different parameters and functions. Work on components made in GaAs has resulted in high-speed electronics and laser diodes at near infrared (~0.7 to ~1 μm window). However, the suitable wavelengths for optical transmission in present communication systems are around 1.3 μm and 1.5 μm (lowest dispersion and lowest absorption in silica fibers, respectively). This motivated research on InP-based material. More recently, GaSb substrates have attracted attention because lasers made in them can operate in the 2 μm to 4 μm region, which includes wavelengths of interest in optical transmission in very low-loss fluoride glass fibers.

In all-optical integrated optics devices, a nonlinear optical interaction is employed to achieve, for example, switches and modulators. This field has grown dramatically during the past decade or so, in parallel with a better understanding of the properties of semiconductor-based microstructures such as quantum wells and superlattices. More recently, there has been a lot of interest in nonlinear polymers, due to their large nonlinearities. Most all-optical devices of current interest are based on the underlying principle of dynamic nonlinear optical effects, by which it is possible to alter the properties of the material (absorption coefficient and index of refraction) by photon absorption near the resonances of the system.

The principles of integrated optical devices, electrooptic, optoelectronic, and all-optical devices have been addressed in other books [4–7], and they will not be discussed here. This book is devoted to integrated optical devices in glass. In particular we will study glass waveguiding components made by an ion-exchange technique.

1.2 GLASS INTEGRATED OPTICS

Glass is an interesting substrate material for integrated optics because of its relatively low cost, excellent transparency, high threshold to optical damage, and availability in substantially large sizes. It is also mechanically very rigid. Furthermore, glass substrates are amorphous, and it is easier to produce polarization-insensitive components in glass. In addition, the refractive index of glasses used in integrated optics (e.g., silicate, phosphate) is close to that of optical fiber and, therefore, coupling losses between the waveguides made in glass and the optical fibers are smaller.

A number of different processes have been employed to make glass waveguides. These processes may be divided into five categories:

1. Sputtering [8, 9].
2. Chemical vapor deposition [10–12].
3. Sol gel coating [13].
4. Ion implantation [14–15].
5. Ion exchange [16].

Waveguides were made by using all these processes, but the ion exchange process has been by far the most popular technique to produce glass integrated optical components.

1.3 ION-EXCHANGED GLASS INTEGRATED OPTICS

In an ion-exchange process, an ion in glass (usually Na^+) is replaced by an ion of a larger size or higher polarizability such as Ag^+, K^+, Cs^+, or Tl^+. Consequently, the refractive index of glass increases locally, giving rise to a waveguide. Ion exchange can be a purely thermal process. However, an electrical field may be applied to accelerate the process. Typically the ions are introduced into glass from a molten salt, but in the case of an electric-field-assisted ion exchange, a thin metallic film has also been used as ion source.

The ion-exchange process is suitable for glass waveguide fabrication for several reasons:

- The process offers considerable flexibility in the choice of fabrication parameters and, therefore, can be optimized for a wide variety of application.
- The process is simple and suitable for high-volume batch processing. The fabricated waveguides are reproducible and have low propagation losses. There is no need for in-situ control of device parameters during the fabrication process due to excellent reproducibility.
- Waveguides with excellent match to conventional single- and multimode fibers can be made, thus minimizing the coupling losses.
- Ion-exchanged waveguides have a planar configuration. This facilitates considerably the use of other materials (e.g., nonlinear polymers) and devices made using other materials (e.g., detectors) with glass-integrated optical devices. Very high-performance hybrid integrated optical circuits can be achieved in this way.

The ion-exchange process has tremendous potential for fabrication of high-performance integrated optics devices. Since fabrication of the first ion-exchanged glass waveguide in 1972 [17], significant progress has been made in this field. Different fabrication processes have been developed.

Passive components such as power dividers-combiners and wavelength multiplexers-demultiplexers have been produced for optical communication systems. Different types of sensors (temperature, refractive index, pressure) have also been demonstrated. Active components such as lasers have been achieved. Very accurate thoretical models have also been developed. It has been shown that these models predict well the behavior of ion-exchanged glass waveguiding devices. In this book we discuss such developments.

1.4 SCOPE OF THE BOOK

This book can be divided into four parts: (1) glass waveguide fabrication processes, (2) glass waveguide modeling and simulation, (3) characterization of glass waveguides, and (4) glass integrated optical devices.

Chapters 2 and 3 explain the processes to make ion-exchanged glass waveguides. The physics and chemistry of ion exchange in glass are discussed in Chapter 2. This chapter explains also the details of fabrication process from salt melts and provides the reader with sufficient experimental data to design and produce waveguides with given optical properties. Chapter 3 concentrates on a special field-assisted ion-exchange process that uses a solid-state silver film as the ion source. This field has seen a lot of progress in recent years. The technique has some unique capabilities that are discussed in Chapter 3.

Chapter 4 examines computational methods to determine index profiles and propagation characteristics of the glass waveguides. One- and two-dimensional solutions (slab and channel waveguides, respectively) are discussed. Different methods (WKB, inverse WKB, effective index, finite difference, finite element, beam propagation) are addressed in this chapter. The results of the refractive index profile, mode profile, and propagation constant calculation using these methods are reported.

Chapter 5 deals with experimental techniques used in the characterization and testing of glass waveguides. It explains the essential techniques to determine basic optical properties of the waveguides, such as their profile and propagation constant, and addresses both slab and channel waveguides.

The optical properties of ion-exchanged waveguides are reviewed in Chapter 6. Straight and curved waveguides are studied. Waveguides with submicron grating, different types of couplers, Mach-Zehnder interferometers, and rare-earth-doped amplifiers and lasers are discussed as well. This chapter also summarizes the most important devices that have been proposed or demonstrated. The optical behavior of these devices and their applications are discussed, too.

REFERENCES

1. Miller, S.E., and L.C. Tillotson, "Optical Transmission Research," *Appl. Opt.*, Vol. 5, 1966, pp. 1538–1549.
2. Kapron, E.P., D.B. Keck, and R.D. Maurer, "Radiation Losses in Glass Optical Waveguides," *Appl. Phys. Lett.*, Vol. 17, 1970, pp. 423–425.
3. Miller, S.E., "Integrated Optics: An Introduction," *Bell Syst. Tech. J.*, Vol. 48, 1969, pp. 2059–2068.
4. Tamir, T., *Integrated Optics*, Springer-Verlag, New York, 1975.
5. Hunsperger, R.G., *Integrated Optics: Theory and Technology*, Springer-Verlag, New York, 1982.
6. Lee, L.D., *Electromagnetic Principles of Integrated Optics*, John Wiley and Sons, New York, 1986.
7. Nishihara, H., M. Haruna, and T. Suhara, *Optical Integrated Circuits*, McGraw-Hill, New York, 1989.
8. Del Giudice, M., F. Bruno, T. Cicinelli, and M. Valli, "Structural and Optical Properties of Silicon Oxynitride on Silicon Planar Waveguides," *Appl. Opt.*, Vol. 29, 1990, pp. 3489–3496.
9. Robertson, G.R.J., and J. Jessop, "Optical Waveguide Laser Using an RF Sputtering Nd:Glass Film," *Appl. Opt.*, Vol. 30, 1991, pp. 276–278.
10. Kominato, T., Y. Ohmori, H. Okazaki, and M. Yasu, "Very Low-Loss GeO_2-doped Silica Waveguides Fabricated by Flame Hydrolysis Deposition Method," *Electron Lett.*, Vol. 26, 1990, pp. 327–328.
11. Kawachi, M., "Silica Waveguides on Silicon and Their Application to Integrated Optic Components," *Opt. Quant. Electron.*, Vol. 22, 1990, pp. 391–416.
12. Hanabusa, M., and Y. Fukuda, "Single-Step Fabrication of Ridge Type Glass Optical Waveguides by Laser Chemical Vapor Deposition," *Appl. Opt.*, Vol. 28, 1989, pp. 11–12.
13. Hewak, D.W., and J.Y. Lit, "Fabrication of Tapers and Lenslike Waveguides by Microcontrolled Dip Coating Procedure," *Appl. Opt.*, Vol. 27, 1988, pp. 4562–4564.
14. Townsend, P.D., "Optical Effects of Ion Implantation," *Repts. on Prog. in Physics*, Vol. 50, 1987, pp. 501–558.
15. Ashley, P.R., and D.K. Thomas, "Low Loss Ion Implanted Ag Waveguides in Glass," Proc. Integrated and Guided Wave Optics Technical Meeting (IGWO'89), 1989, pp. 152–155.
16. Over 100 papers have been published. A fairly complete list of these papers up to 1989 is given in L. Ross, "Integrated Optical Components in Substrate Glasses," *Glastech. Ber.*, Vol. 62, 1989, pp. 285–297. Some of these papers as well as more recent works will be addressed in the coming chapters.
17. Izawa, T., and H. Nakagome, "Silver Ion-Exchanged Glass Waveguides," *Appl. Phys. Lett.*, Vol. 21, 1972, pp. 584–586.

Chapter 2

Ion Exchange from Salt Melts

Jacques Albert

Communications Research Centre, Communications Canada, Ottawa

2.1 INTRODUCTION

This chapter describes optical waveguide formation in glass by ion exchange from salt melts. The physical and chemical processes peculiar to this type of exchange are detailed in Section 2.2. Section 2.3 looks at the experimental conditions and methods to form optical waveguides. Finally, the properties of the waveguides thus formed are given in Section 2.4.

    The references listed at the end of the chapter are not intended as a complete survey of the existing body of literature on the subject. They are included to guide the reader to further reading on topics that must be covered quickly due to space limitations.

2.1.1 Purpose and History of Ion Exchange from Melts

In multicomponent oxide glasses, the various constituents may be classified according to the bond strength between their cations and the oxygen atoms. Oxides with strong bonds are called *network formers*. These include SiO_2, B_2O_3, GeO_2, P_2O_5, and As_2O_3. Glasses with large relative amounts of network formers have higher glass transition temperatures and, in general, higher viscosities at all temperatures. There are also network intermediates, which contribute to the strength of the network but cannot form glasses by themselves; ZnO and PbO belong to this class. Finally, there are network modifiers with relatively loose bonds to the remainder of the network. These are generally added to the glass to give it

---

some desired property, such as a lower softening point or a greater resistance to bubble formation during melting. Some of the more common additives are Na_2O, CaO, K_2O.

Under certain conditions, described in detail in this chapter, it is possible to replace some of the network modifier ions by others with the same valence and chemical properties. In the exchanged region thus formed, various glass properties are modified. Of interest from a historical perspective is the strengthening of glass surfaces, which has become a standard industrial process [1]. In this case, also referred to as *ion stuffing*, the ions introduced are larger than those coming out of the glass. At the temperature of exchange, the glass volume can adjust to accommodate the larger ions but, as the glass cools, large compressive surface stresses develop. These surface stresses tend to strengthen the glass by inhibiting microcrack formation and propagation.

Of course, ion exchange also modifies the electrical and optical properties of the glass. The new ions have different polarizabilities, sizes, and mobilities. This fact is used to fabricate optical waveguides by increasing the refractive index in selected areas of a glass substrate [2]. The exact mechanisms by which the index increases are also discussed in detail later. Optical waveguide fabrication from ion exchange dates back to the early 1970s [3, 4]. However, the real impulse came in the 1980s with the realization that optical fiber networks would become widely used in all kinds of applications (communications, sensors, light distribution, *et cetera*). With these advances, a huge market developed for low-loss, low-cost, passive components for all kinds of signal-handling functions between fiber links [5]. At this point various companies around the world (in all cases, glass manufacturers) linked up with research programs already in progress, mainly in universities and government, to develop products. Figure 2.1 shows schematically the process required to fabricate a channel optical waveguide in a planar substrate from a melt. After cleaning, a masking layer is deposited on the surface and patterned photolithographically. The whole surface is then exposed to a salt melt containing the ions to be introduced in the glass, in a temperature controlled furnace. Upon removal from the furnace, residual salts and the masks are removed, leaving a bare glass surface in which an ion-exchanged channel has been formed. It is the purpose of this chapter to describe in detail all the processes involved.

Although it is relatively easy to fabricate ion-exchange waveguides, it takes a good control over the physics and chemistry of the process to achieve consistently low-loss (<0.1 dB/cm) waveguides of accurate dimensions and shape for matching to optical fibers. It is not surprising, therefore, that in most cases specially designed glasses have been developed by these manufacturers for their ion exchange work.

Multimode components such as splitters appeared first, followed more recently by single-mode optical circuits of increasing sophistication. Applications involving active components are still at the research level.

FABRICATION STEPS

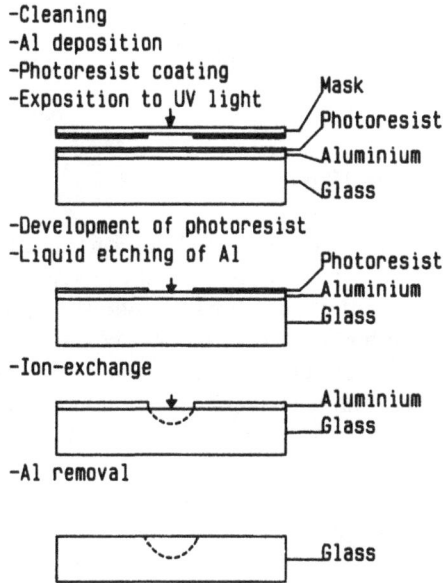

-Cleaning
-Al deposition
-Photoresist coating
-Exposition to UV light

-Development of photoresist
-Liquid etching of Al

-Ion-exchange

-Al removal

Figure 2.1 Step-by-step procedure for channel waveguide fabrication by ion exchange, here the mask material is aluminum [6].

2.1.2 General State of the Art (as of 1991)

- The maximum index of refraction of the exchanged layers lies between 0.001 and 0.16 above that of the substrate. This allows the fabrication of multimode guides with large numerical apertures and also single-mode guides with good reproducibility and control.
- The best propagation-loss results are obtained with buried single-mode waveguides and are lower than 0.1 dB/cm, where loss measurements become more difficult.
- Mode matching to single-mode fibers yields between 0.1 and 0.3 dB of loss per interface. Matching losses below 0.1 dB are predicted for optimized processes simulated using modeling [7].
- The maximum size is usually not limited. Exchange depths of millimeters have been attained (for purposes other than waveguiding). However, for certain ion-glass combinations it is difficult or impossible to exchange deeper than about 10 μm.

2.2 PHYSICS AND CHEMISTRY OF THE ION-EXCHANGE PROCESS

This section describes the details of the exchange at the microscopic level. A basic knowledge of this subject is necessary to design ion-exchange processes for specific purposes and to understand the properties of the waveguides obtained under different conditions.

2.2.1 Ionic Conductivity in Glass

As mentioned in Section 2.1.1, network modifier ions are not attached very tightly to the silicate network; they are ionically bound to oxygen atoms. They also have a temperature-dependent mobility $\mu(T)$, which follows an Arrhenius-type behavior ($\mu = \mu_0 \exp(-Q/RT)$, with Q an activation energy and R the gas constant), being essentially diffusion driven. Under appropriate conditions they will move readily, especially at higher temperatures (typically a few hundred degrees Celsius). Such conditions are met when a concentration gradient of similar ions is created (exposing the surface of the glass to a melt solution containing other ions with similar chemical properties, for instance) or when ions from a source are driven in the glass with the help of an electric field. Note that the oxygen ions to which the alkalis are bound are much more tightly fixed in the network. They form covalent bonds with silicon atoms. Finally, because the mobile species are charged, the exchange takes place on a one-to-one basis to preserve glass neutrality.

2.2.2 Driving Mechanisms for Ion Exchange

Two basic forces can drive the exchange: the thermal agitation and nonzero mobility of certain ions in the glass at sufficiently high temperatures, and a potential difference set up across the glass causing a current of ions to flow. The thermal force may be used alone and is sufficient to fabricate optical waveguides in many instances. When an electrical field is used, the thermal contribution is also present but its influence can become negligible if the electric field is large. Roughly speaking, field-assisted exchange from melts is used for large (multimode) waveguides with relatively sharp refractive-index boundaries or to bury waveguides under the surface of the substrate while thermal processes are used to define small monomode surface guides. Details of the two processes follow.

2.2.2.1 Thermal

Qualitatively, thermal ion exchange proceeds as follows. A glass substrate containing A^+ ions is immersed in a molten salt containing chemically similar ions, called B^+ here, for example. At the glass-melt interface, both ion concentrations initially drop suddenly from finite values to 0 (see Figure 2.2). This is clearly a nonequilibrium situation as B^+ and A^+ ions are almost perfectly interchangeable in both the melt and the glass. Therefore, thermal agitation at the interface produces random collisions in which one B^+ ion replaces one A^+ ion, and this process gradually diffuses away from the interface. Of course, in the melt, the A^+ ions move much more rapidly away from the surface (and are "lost" in what can be considered an infinite reservoir of B^+ ions) than the B^+ ions in the glass, which slowly invade a very thin layer near the surface of the substrate. The process accelerates at higher temperatures because of stronger thermal agitation and also because the glass matrix, through which these ionic motions take place, is less rigid.

When the glass is lifted out of the melt, but kept at a high temperature, the exchange continues without a supply of new B^+ ions. The result is that the ions,

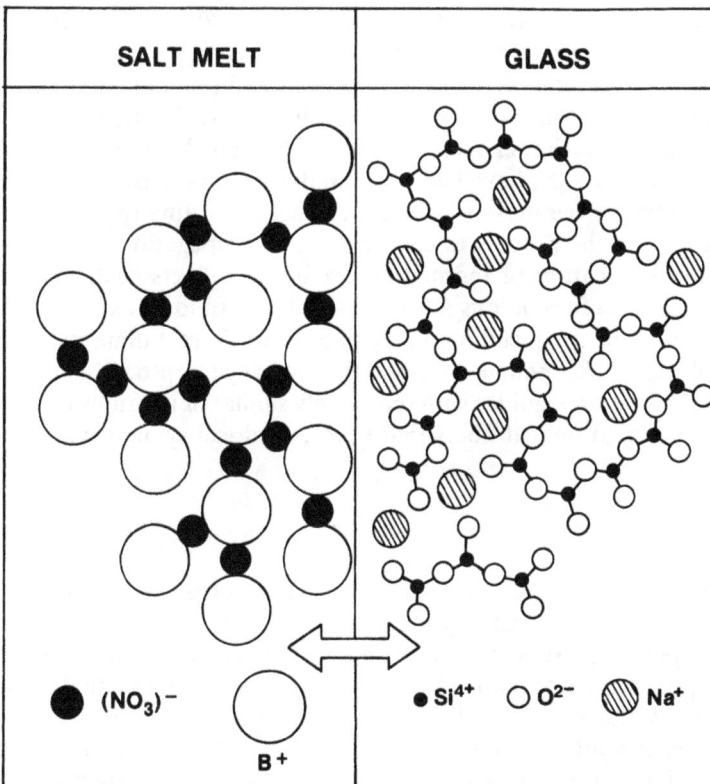

Figure 2.2 Diagram of melt-glass interface (here $A^+ = Na^+$).

which are already in the glass, will tend to redistribute themselves to reach equilibrium (i.e., uniform concentration of both B^+ and A^+ throughout the substrate) by moving in deeper but with decreasing surface concentration. The process stops (or rather becomes infinitely slow) only when the source of heat is removed and the substrate allowed to cool to room temperature. Typical exchange temperatures range from 200°C to 550°C and generally do not lie too much above the melting point of the salts used as sources of exchanging ions because excessive heat may lead to damage of the surface of the substrate due to nitrate decomposition and excessive thermal relaxation of the glass. The resulting concentration profile has a maximum at the surface and decreases monotonically inside the substrate because of the configuration and nature of the process.

Channel waveguides are defined by masking part of the substrate prior to immersion in the melt. The mask material must not be susceptible to attack by the harsh alkali melts usually employed in ion exchange and must be impermeable to ionic motion so as to block the exchange. Various materials have been used but a convenient choice is evaporated aluminum, at least 100 nm thick. In some cases, nonmetallic masks are preferred. In silver-sodium exchange for instance, silver ions tend to reduce to metallic silver at the boundaries of metallic masks, increasing the absorption losses of the waveguides. In such cases, dielectric masking materials may be used, as long as they can tolerate the harsh salt melt environment. Anodized aluminum, SiO_2, and Si_3N_4 have been used in this context. However, it is desirable to be able to remove the mask selectively (without affecting the underlying glass) after completion of the process to ensure low-loss propagation in the waveguide.

The isotropic nature of thermal ion exchange reflects itself in the fact that the concentration profiles of ions B^+ in the substrates (and by extension the refractive index profiles of the resulting waveguides) have lateral dimensions exceeding the width of the mask opening by twice the exchange depth. This makes it very difficult to fabricate waveguides with reasonably similar dimensions in all directions for good matching to optical fibers that have cylindrical symmetry.

2.2.2.2 Field Assisted

In the field-assisted case, a potential difference is set up between two melts on either side of a glass substrate by means of immersed electrodes (an alternative consists of depositing a solid electrode on the cathode side of the substrate). Contrary to the previous case, both A^+ and B^+ ions move in the same direction in the glass, driven by the potential difference across it. Mobile ions become charge carriers for the current that flows between the electrodes. For each B^+ ion coming in from one side, an A^+ comes out on the other side. The integrated current is used to monitor the total amount of dopant ions introduced in the glass.

Of course, even in the field-assisted case, the temperature must be high enough for the ions to have sufficient mobility. Therefore, there will be a thermal contribution to the exchange in addition to the field-assisted component. By adjusting the driving voltage and temperature, the relative contributions of the two driving forces can be varied. This allows some control over the profile of index change, which is quite different in the two cases.

Channel waveguide fabrication is similar to the thermal case except that the lateral spreading is controlled differently. Here, the ions follow electric-field lines, linking the source and sink of the potential difference set up across the substrate (Figure 2.3). For planar guides the field lines are straight but for exchange through a narrow opening there is a significant spreading of the field lines. This leads again to waveguides much wider than deep and problems of mode matching to circular fibers. Other types of field-assisted exchange are discussed elsewhere in this book.

2.2.3 Materials and Conditions for Ion Exchange from Melts

The ion-exchange process depends greatly on the materials and conditions used. Only certain types of glasses may be used, and a limited number of ions have the necessary properties to participate in the exchange.

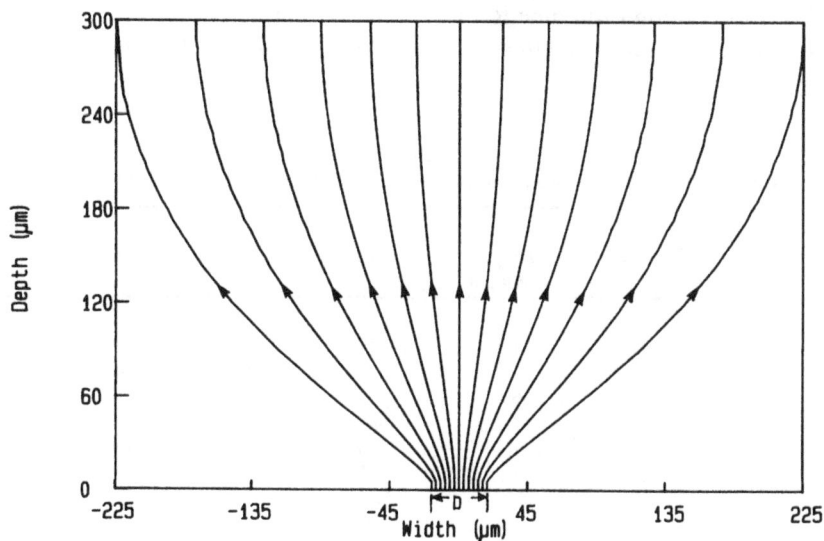

Figure 2.3 Electric field lines in glass for exchange through a narrow opening of width D, equal to 36 μm here [10].

2.2.3.1 Glasses and Dopant Ions

As was mentioned in Section 2.1.1, specially designed glasses are used in commercial processes. However, there are several types of widely available glass substrates in which ion exchange has produced high-quality waveguides for research. The composition of some of these glasses is listed in Table 2.1.

First and foremost is the simple microscope slide, available from several manufacturers at very low cost. It consists of soda lime glass, one of the most widely used materials because of its ease of handling in the manufacture of mass-produced everyday objects (windows, drinking glasses, bottles). The key to its success as a substrate for ion exchange is its high sodium content. Sodium ions are among the most mobile in silicate glasses and are easily replaced by ion exchange. Also, surface scattering losses do not pose problems for waveguiding applications because these slides are usually fire polished. This yields a surface that is very smooth, although not necessarily flat (by optical standards). The limit to the usefulness of soda lime glass in applications with stringent requirements comes from the presence of metallic impurities (such as iron oxides), which cause volume optical absorption. Finally, a potential drawback to using microscope slides is the possible variation in properties with time. Manufacturers do not specify important parameters such as refractive index, birefringence, or homogeneity for these products; therefore, they may change from batch to batch. Also manufacturers may change the glass formulation from time to time, yielding completely different ion exchange characteristics. In spite of all this, we have used microscope slides (always the same

Table 2.1

Glass Constituents and Annealing Temperature of Common Substrates for Ion Exchange

Oxide Content in Weight %	Typical Soda Lime	Fisher Premium	Corning 0211	Schott BK7	Pyrex Corning 7740	Bausch and Lomb 2046
SiO_2	71–74	72.2	65	69.6	81	67
Na_2O	13–15	14.3	7	8.4	4	25.6
K_2O	0.1–2	1.2	7	8.4	—	—
CaO	6–11	6.4	—	—	—	—
MgO	3–5	1.2	—	—	—	—
B_2O_3	—	—	9	9.9	13	—
Al_2O_3	0.1–1.2	1.2	2	—	2	7.4
BaO	—	—	—	2.5	—	—
ZnO	—	—	7	—	—	—
TiO_2	—	—	3	—	—	—
Traces	0–2	0.33	—	1.2	—	—
T_{ann} (°C)	545–555	near 550	550	547	565	507

brand) for various exploratory projects over a period of over five years with excellent reproducibility (at the level of the measurement accuracy).

For higher-quality waveguides, an optical glass is required. Some of the most commonly used are borosilicates from Schott (BK7) and Corning (0211). Again, a high sodium content allows ion exchange to proceed easily and relatively quickly. There is also a sizeable quantity of potassium, another good exchanger but with much smaller mobility. Substrates are available in various shapes with at least one face polished to optical quality standards of smoothness and flatness. These glasses are nominally transparent between 400 and 1700 nm.

There is a variety of dopant ions to choose from, but they have unique characteristics that determine the applications where each is most useful. Having said that sodium-rich glasses are among the most used because of the high sodium mobility, the dopant ions for these glasses have to be monovalent anions with similar chemical properties. Obviously, other alkali metals exchange with sodium: Li, K, Cs, Rb. In addition, much work has been done with Ag and Tl, and there have been reports of exchange with Cu and Sn. The properties of waveguides obtained with the most common of these exchangers will be described in Section 2.4.3.

2.2.3.2 Temperatures

The temperature range over which the exchange must take place is bounded from below by the melting point of the salts used and from above by the temperature sensitivity of the various components involved. The typical multicomponent glasses have a softening point near 700°C, at which they readily deform under their own weight. Their annealing point lies in the vicinity of 550°C (glass manufacturers warn that if glass is heated to a temperature equal or higher than the annealing temperature minus 200°C, deformations in the precision finished surfaces and slight changes in refractive index may occur). Table 2.2 shows the melting temperature of the most commonly used salts. We have found that the most reliable temperatures are found at about 40 to 60°C above the melting point of the salt, to ensure high mobility of ions in the melt while avoiding problems associated with high temperatures. When nitrate salts are used, concerns have been raised over chemical attack of the glass surface by NO_3 ions above 400°C.

2.2.3.3 Relative Ion Sizes and Substrate Composition

All of the ions listed previously have different sizes. It follows that they have different mobilities in the glass network, that their maximum concentration in the glass may be different, and that they leave the glass in a different state of stress

Table 2.2
Common Salts Used in Ion Exchange, with Their Melting Points

Salt	Melting Point (°C)
AgNO₃	212
AgCl	455
NaNO₃	307
KNO₃	334
KNO₃–AgNO₃ (37:63 mole %)	132
Li₂SO₄–K₂SO₄	515
KNO₃–NaNO₃ (50:50 mole %)	220
KNO₃–Ca(NO₃)₂ (34:66 mole %)	150
TlNO₃	206
CsNO₃	414
CsCl	646
CsNO₃–CsCl	405
RbNO₃	310

after the exchange. Lithium ions are smaller than sodium ions and induce destructive tensile stresses for instance. The effect of different ion sizes will be discussed in detail in Section 2.4.1.

In certain cases, especially in glasses with more than one exchangeable species, a multicomponent melt is needed to maintain the structure of the substrate's glass forming network during the exchange. This is especially true for multistep processes involving back diffusion (see Section 2.3.3).

2.3 EXPERIMENTAL CONSIDERATIONS AND PROCESS PARAMETERS

There are as many ways to fabricate optical waveguides from melts as there are laboratories doing it. What is described here is a basic setup that can be used for the reliable fabrication of experimental devices. Of course, mass production in an industrial environment would require different conditions.

2.3.1 Equipment

Fortunately, most of the equipment needed for ion-exchange work is relatively inexpensive. Basically, all that is needed is a furnace capable of reaching temperatures where ions become mobile (200–500°C is usually sufficient), glass substrates, and salts. The other items needed are usually stocked in most chemistry laboratories (acids and solvents for cleaning, ultrasonic baths, *et cetera*).

2.3.1.1 Furnace, Crucible, Salts

Although tube furnaces may be used, a more convenient approach uses a crucible furnace with a vertical chamber accessible from above. It is then easier to control the insertion and removal of the substrate from the melt, as well as to introduce various additional elements such as stirrers and thermometers. For most purposes the furnace should be able to cover the range from 100–600°C. A holding device may be suspended from the top cover to hold the substrate above the melt while it reaches the desired process temperature and to control its insertion and removal from the exchange bath (Figure 2.4).

Most of the ions used in ion-exchange work may be obtained from salt melts. Nitrates of various kinds are often used. The salts must be of the highest available purity. They are simply placed in a crucible in their solid form without further

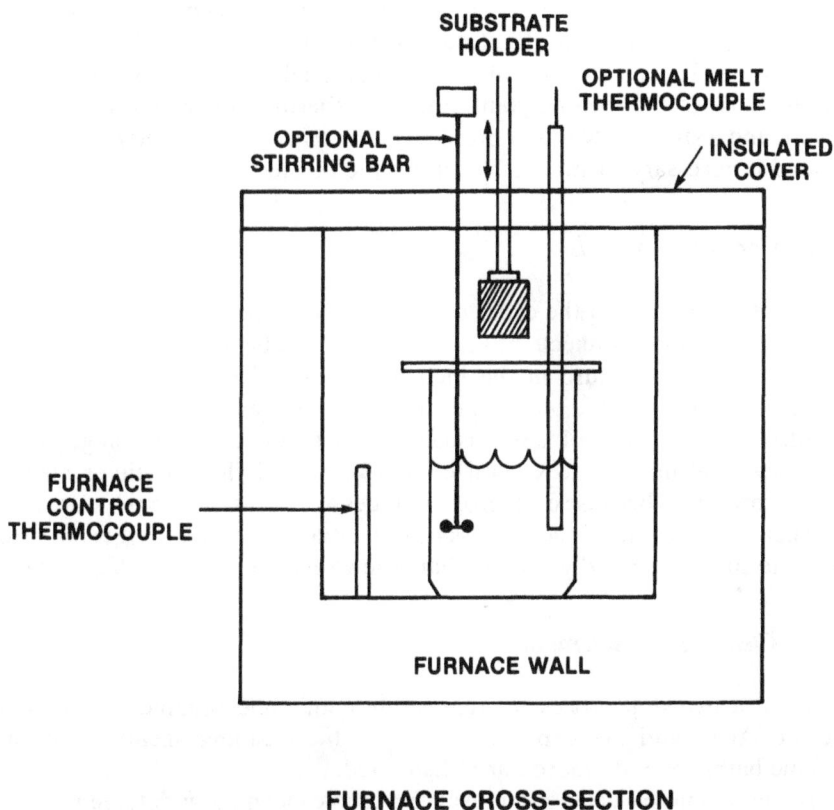

FURNACE CROSS-SECTION

Figure 2.4 Furnace and exchange apparatus cross section.

treatment. They may be mixed at this stage to form diluted melts. Each salt has a definite melting point but mixtures are less predictable. In fact some mixtures have been used specifically to lower the melting point of the salts and, by extension, the minimum temperature of exchange. Multicomponent melts are also used to provide more control over the index profile and state of stress of exchanged layers, and it is even possible to supply controlled amounts of dopant ions to a melt while the exchange is taking place (by electrolytic release) [2].

The crucible should not contaminate or otherwise corrupt the melt. We have found stainless steel to be an acceptable material for applications not exceeding 450°C. Alumina crucibles have been tried but tended to break occasionally.

The fact that the process takes place in a liquid (a volume of approximately 150 ml in our case) smoothes out any temperature gradient along the length of the substrate (because of convection currents arising from slightly nonuniform heating) and also any temporal temperature fluctuations. There is a debate in the ion-exchange literature over the use of stirred melts to ensure a constant concentration of exchanging ions at the surface of the substrate by removal of the outgoing ions from that area. This may be a problem for small melts or when exchange temperatures are close to the melting point of the salt (because ionic motion is then very sluggish), and a stirrer solves that problem easily. However, in other cases we have not found it necessary to stir the melt from the outside.

2.3.1.2 Control Apparatus

A temperature control on the order of 1°C is usually provided with good furnaces but the thermocouple connected to the controller sits beside the crucible and may not reflect the temperature in the melt accurately. For high-precision work, a separate thermocouple in a glass tube should be inserted in the melt and used by the controller. Time control is not usually a problem because exchange durations last from several minutes to several hours. Reproducibility problems may occur with short processes because exchange continues as long as the temperature remains high. Therefore the end of the process is difficult to determine, as it is not possible to cool the substrate rapidly (in less than a few seconds) without damaging it.

2.3.1.3 Cleaning Considerations

In general, a two-step process to remove inorganic and organic contaminants is sufficient. Acids and overexposure to any of the cleaning agents, especially in ultrasonic baths, may do more harm than good.

For inorganic particulate removal, ultrasonic cleaning in detergent, followed by rinsing in *deionized* water is sufficient (ions present in untreated water may combine with surface molecules of the glass to form a weaker layer). For organic matter cleaning, isopropyl vapors have proven very effective in cleaning glass.

Chlorine-based solvents are less successful in yielding clean surfaces, possibly because of a reaction with adsorbed water to form hydrochloric acid. The acid would leach alkali from the surface of the glass and destroy it partially.

To achieve the best results, no liquid droplets should be left to dry on the surface after any of the cleaning processes. The water left after the rinsing can be blown off with a filtered jet of dry gas, and the removal from the vapor degreaser should be slow enough for any liquid alcohol on the surface to evaporate.

2.3.1.4 Special Requirements for Field-Assisted Exchange

In field-assisted exchange with salt melts, a large potential difference (typically hundreds of volts) is applied between two sides of a substrate, at least one of which must be in contact with the liquid melt. The problem is to avoid leakage of the conducting melt, which could cause a short circuit. Special cell designs are needed for this purpose. The cell must be able to contain the melt(s) securely while temperatures vary by hundreds of degrees. Two possible cell designs are shown in Figure 2.5. Platinum wires are used as electrodes for immersion in the conducting melts.

2.3.2 Process Sequence

The salts, solid at room temperature, are placed in the crucible and in the furnace. The cleaned substrate is also placed in the furnace, outside the melt. Once the desired process temperature is reached and stable, the substrate is immersed in the melt and exchange begins. This event can be timed to within a few seconds. For accurate time control of the exchange, the substrate must be removed from the melt and cooled rapidly. This is achieved by removing the substrate completely from the furnace and allowing it to cool in air. A layer of molten salt usually sticks to the surface and solidifies in about 10–15 seconds. This indicates that the exchange is stopped for all practical purposes because the mobility decreases exponentially with temperature. A more rapid cooling results in damage (catastrophic) to the substrate. The crystallized salts can be removed by rinsing in water when the substrate has reached room temperature. The procedure for channel waveguide fabrication differs only in the sense that a mask must be patterned photolithographically on the substrate before the exchange [6] (masks were described in Section 2.2.2.1).

2.3.3 Multistep Processes

Burial of the waveguide mode is a desirable goal for glass waveguides. Because these are generally used as passive components, access to the guided optical energy

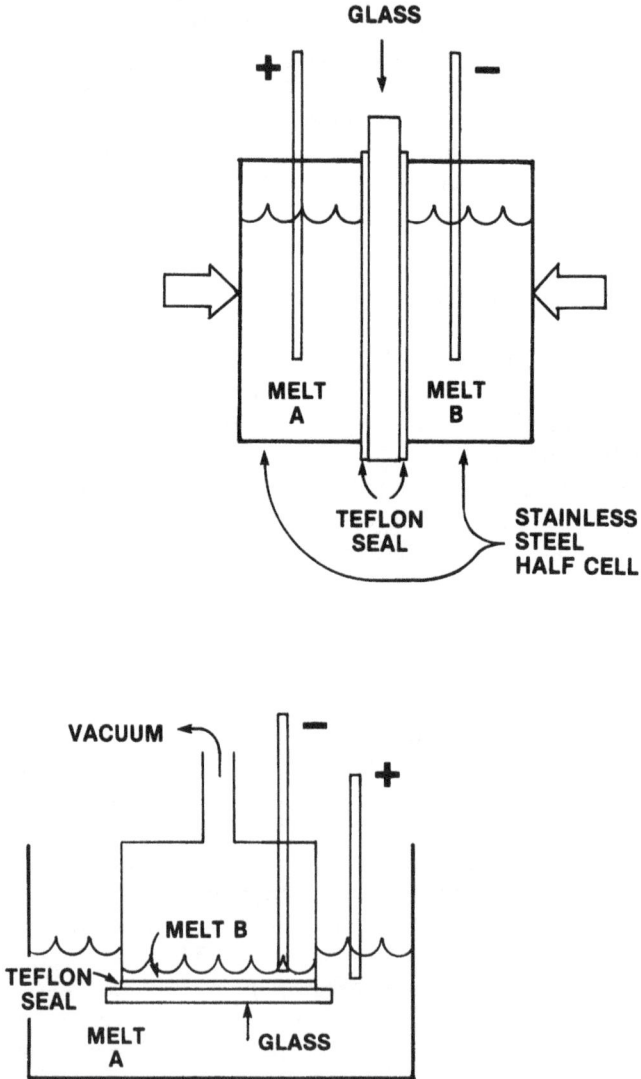

Figure 2.5 Cell designs for field-assisted exchange. (Adapted from [3 and 9].)

from the surface of the substrate is not usually necessary. At the same time, a buried layer reduces the scattering losses associated with surface roughness and improves the mode matching of the waveguides to optical fibers.

One approach for waveguide burial, compatible with the ion exchange process, consists of a multistep exchange. In this case, the ions driven out of the glass in the first exchange are reintroduced from the surface in a subsequent exchange. It is not necessarily as trivial as it sounds, because the glass properties have changed drastically in the exchanged layer. The second exchange may yield tensile stresses causing damage to the surface or prove to be much too slow to be practical. In some cases these problems can be reduced by using multicomponent melts that "fill out" or reconstruct the glass surface better after the initial exchange. Much of this type of work still involves trial and error because the kinetics of the ion exchange in multicomponent glasses is a poorly understood topic.

Instead of burying the actual waveguide core below the surface, another approach may be used to symmetrize and bury the *mode*. By starting from an appropriate profile, a simple annealing procedure redistributes the dopant ions over a larger, more circular cross section, and the resulting optical mode matches very well that of standard optical fibers. The optical power at the glass-air interface is also reduced somewhat. This annealing is done simply by introducing the substrate in the heated furnace after the initial exchange. It is better to cool and clean the surface between the first exchange and the annealing step because the residual salt left on the surface is nonuniform.

In another kind of multistep process, a first exchange with an ion that has relatively low mobility is used as a mask for a second exchange with a mobile ion. Potassium is almost immobile at the low temperatures used in silver exchange, for instance. Of course, the initial mask used is then the negative version of the mask needed for a conventional process. This solves the problem of lateral diffusion under the edges of the conventional surface masks used and allows for more confinement of the optical energy. An extension of this involves using one ion exchange to define a waveguide pattern while another one defines a structure on it (such as grating, or lens), again by using two processes at different temperatures. Figure 2.6 shows an example of a multistep process to fabricate a buried waveguide with cylindrical symmetry (obtained by numerical simulation) [10].

2.3.4 Postprocessing of Waveguides

Being controlled by a diffusion process, ion-exchanged waveguides lend themselves to various forms of heat-induced modifications. In particular, mode transformation tapers can be formed by nonuniform heating of a channel waveguide. Because the local concentration of ions (and therefore the refractive index) diminishes at the same time as the size increases, the tapers thus formed are naturally adiabatic in the sense that the mode index is roughly conserved along the taper. Another tapering method consists of lowering the masked substrate into the melt gradually,

(a)

(b)

(c)

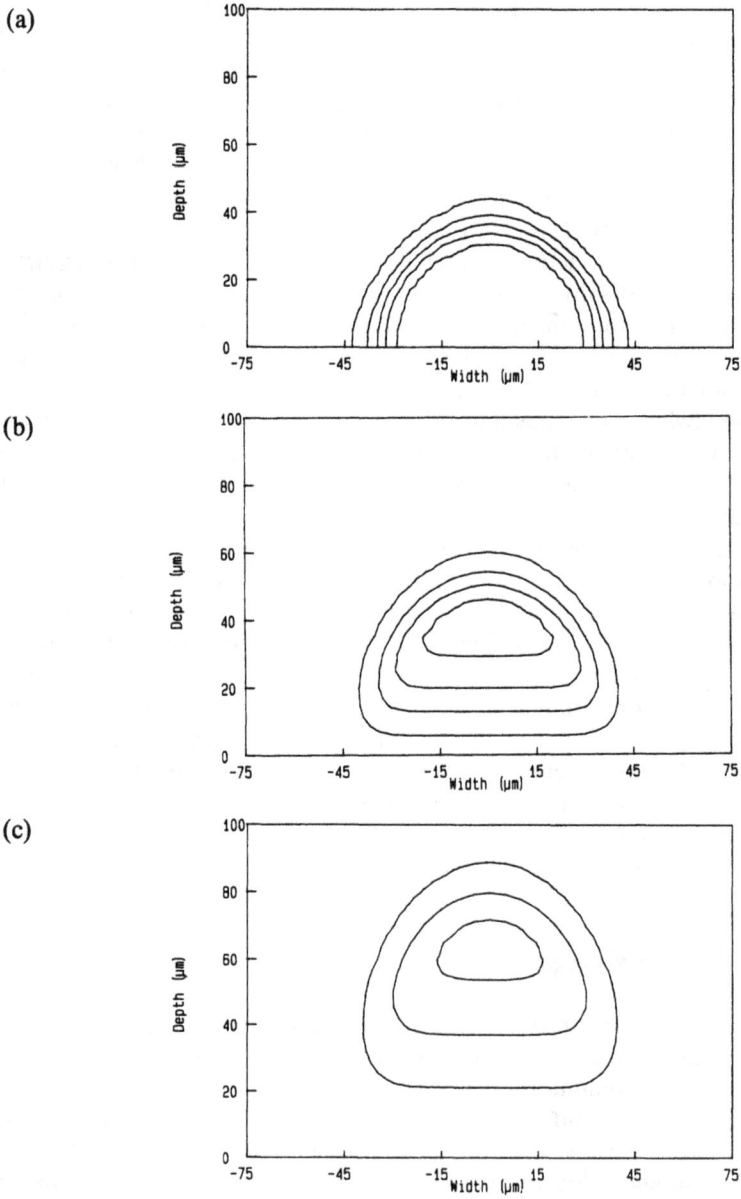

Figure 2.6 Numerical simulation of multistep cylindrical waveguide formation [10].

(d)

(e)

(f)

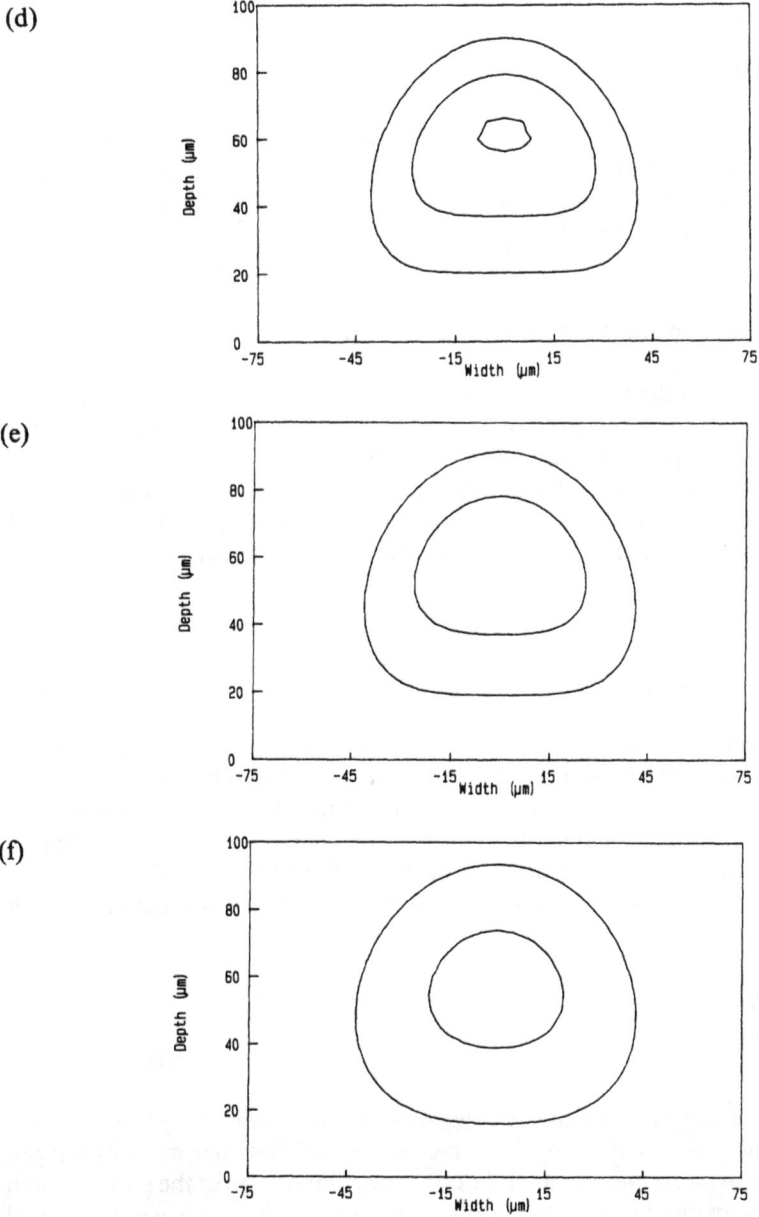

Figure 2.6 Continued.

so that the exchange duration (and therefore waveguide size) decreases monotonically.

2.4 WAVEGUIDE PROPERTIES

Quite a wide range of optical waveguide properties can be achieved through ion exchange. Both tightly confined and weakly guiding single-mode guides are possible, as well as large multimode waveguides with various numerical apertures. Surface and buried guides have also been demonstrated.

2.4.1 Index Change *versus* Concentration of Dopant Ions

Throughout the ion-exchange literature, the change in refractive index is taken to be proportional to the concentration of dopant ions introduced in the glass. Various models have been proposed to explain this fact. They all rely on two mechanisms for the relationship between exchange and index change: the difference in polarizability per unit volume of the ions involved, and the different states of stress of the glass before and after the exchange. The detailed relationships are described in the following text.

2.4.1.1 Chemical Change

The usual explanation for the index change resulting from ion exchange is based on the fact that the ions participating in the exchange have different electronic polarizabilities and that they occupy a different volume of the glass [11]. Quantitatively, a very accurate empirical model exists to predict the value of the index change that results from replacing one ion by another in the bulk composition of the glass.

First, the refractive index n is expressed in terms of the refraction per unit volume:

$$n = 1 + \frac{R_0}{V_0} \tag{2.1}$$

where R_0 is the refraction per mole of oxygen atom (R_0 is defined by this equation and has the same units as V_0), and V_0 is the volume of glass per mole of oxygen atoms. Then, these two quantities are related to the composition of the glass through the following linear relationships (where the constants C, b_i, and a_i were obtained experimentally):

$$V_0 = C + \sum b_M N_M \tag{2.2}$$

$$R_0 = \sum a_M N_M \tag{2.3}$$

In these equations, N_M is the number of moles of ion I contributed by the molecular component M, of chemical composition $I_m O_n$ (K_2O or Al_2O_3, for instance) per mole of oxygen ions contributed by all the components of the glass. It is calculated from the composition data in weight fraction by the following formula:

$$N_M = \frac{m_M f_M / W_M}{\sum n_M f_M / W_M} \tag{2.4}$$

where W_M is the molecular weight (weight of 1 mole) of the component M, f_M is its weight fraction in the glass, m_M is the number of ions I, and n_M is the number of oxygen ions in the molecular formula. Relevant coefficients are listed in Table 2.3. We seek the effect of replacing a fraction χ of ions A^+ by ions B^+. A simple algebraic manipulation of equations (2.2) and (2.3) yields

$$\Delta V_0 = \chi \Delta V = \chi N_A (b_B - b_A) \tag{2.5}$$

$$\Delta R_0 = \chi \Delta R = \chi N_A (a_B - a_A) \tag{2.6}$$

Neglecting terms of order 2 in Δ, we get for the index change

$$\Delta n \approx \frac{\chi}{V_0} \left(\Delta R - \frac{R_0}{V_0} \Delta V \right) \tag{2.7}$$

It is obvious to see from this model that a linear relationship exists between the concentration of new ions in the glass (as expressed by χ) and the index change.

This model for the refractive index holds very well for bulk changes in composition. However, in the case of ion exchange over thin layers at the surface of a much thicker substrate, another effect must be taken into account. The volume change may not be allowed because the rigid substrate strongly resists bending to accommodate this localized expansion or compression. Stresses develop instead, and the value of ΔV calculated previously is not accurate. The amount of stress relief due to bending or stretching the whole substrate is on the order of the ratio of the thickness of the exchanged layer over that of the substrate.

One important case where this is evident is the potassium-sodium exchange in soda lime glass. The equations given earlier predict an index change that is two orders of magnitude smaller than measured values on thin (μm sized) layers [12]. Exchange over thick layers (mm sized), however, yield the predicted index change,

Table 2.3
Coefficients for Calculating the Volume and Refraction of Silicate Glasses
as a Function of Composition. (Adapted from [11])

Ion	W (g/mole)	a	b
Li$^+$	29.89	4.60	3.9
Na$^+$	61.98	6.02	8.7
K$^+$	94.20	9.54	15.5
Rb$^+$	186.94	12.44	22.0
Cs$^+$	281.81	17.48	30.3
Tl$^+$	424.74	31.43	23.0
Ag$^+$	231.74	15.97	12.72

confirming that the model failure is related to a surface effect. We show in the next section how to take surface stresses into consideration in the analysis.

2.4.1.2 Stresses

Surface stresses in ion exchange arise from the fact that, at the low temperatures at which the process takes place (350–400°C for potassium), the glass is well below its strain point (510°C) and the surface is prevented from expanding or compressing laterally by the resistance to bending of the relatively thick substrate. Therefore, the volume change used in equation (2.7) is generally inaccurate. In the more important case of volume expansion (exchanges that result in decreased volumes yield fragile or damaged layers because of the tensile stresses induced), the only direction of free expansion is normal to the surface and results in a swelling of the glass. An example of such swelling is shown in Figure 2.7 as a height discontinuity at the boundary of the exchanged area [12].

To account for these considerations in equation (2.7), we apply a volume correction obtained by imposing surface stresses to the value of ΔV calculated previously. The state of stress at the surface is uniform and isotropic in the plane of the surface and zero in the normal direction. The stress tensor components are

$$\sigma_y = \sigma_z = \sigma_0; \qquad \sigma_x = 0 \tag{2.8}$$

with the x component taken along the normal direction. From the theory of elasticity, these stresses may be related to a volume correction by

$$\frac{\Delta V}{V} = 2\frac{\sigma_0}{E}(1 - 2v) \tag{2.9}$$

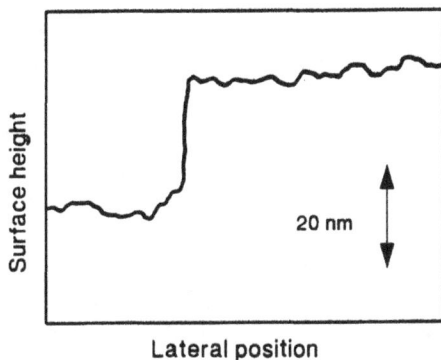

Figure 2.7 Height discontinuity at the boundary of an exchanged layer, showing volume increase (measured with a stylus profilometer) [12].

where E is Young's modulus and v is Poisson's ratio. These values are available from manufacturers. For soda lime glasses they are

$$E = 7 \times 10^4 \text{ Newton/mm}^2 \qquad v = 0.2$$

The main unknown in this analysis is the magnitude of the surface stress σ_0, and an independent measurement is needed. The usual measurement methods relying on measuring the change in birefringence of the samples, by either polarimetry or mode spectroscopy of orthogonally polarized modes, fail in this case for the following reason. The tabulated strain-optic tensor components used are quite sensitive to the glass composition, and these values for the substrate cannot be expected to be reliable in the exchanged layer, where the glass composition is completely different. Because this layer is precisely where the stress-birefringence relationship is needed, the method fails.

An independent method consists of performing an exchange in a very thin sample that does yield to the surface stresses by bending. A measurement of this bending allows the calculation of the surface stresses by elastic analysis, the coefficients of which depend much less on the exact glass composition. We can estimate σ_0 to lie between -700 and -1000 N/mm^2 for potassium-sodium exchange in soda lime glass [1, 13]. Of course, the exact value depends on a number of factors, such as the glass composition and the temperature of the process. Lower process temperatures should result in higher stress because of the reduced viscosity. Taking the highest value of stress to compare the calculation with our index change data at a relatively low temperature (350–400°C), we obtain a volume correction to ΔV that brings the value of Δn obtained with (2.7) in agreement with experimental results (i.e., about 0.009). In addition, the stress analysis can also be used to explain the birefringence observed in some types of waveguides (for potassium-sodium exchange, the index change for TM polarized modes is systematically about 0.002 higher than for TE modes).

The birefringence of anisotropically stressed glass is given by

$$\delta = \Delta n_{TE} - \Delta n_{TM} = B\sigma_0 \qquad (2.10)$$

where B is the birefringence factor of the material (for soda lime glass $B \approx 2.4 \times 10^{-6}$ mm^2/Newton). This gives a δ value of 2.4×10^{-3}, somewhat higher but still close to the measured values. The same calculations were performed for the case of silver-sodium exchange, and the results are summarized in Table 2.4.

It is important to note here that the general linear relationship between index change and concentration of new ions is preserved in the combined model, including polarizability changes and stresses if we take σ_0 to be proportional to χ.

Table 2.4
Results of Modeling the Stress-Induced Contribution to the Change in Index
for Silver-Sodium and Potassium-Sodium Exchanges [12]

Ions	ΔV	ΔR	Δn	$\Delta V'$	$\Delta n(+$ str.)	Δn(exp.)	δn	δn(exp.)
K$^+$-Na$^+$	1.054	0.546	0.0003	-0.25	0.0089	.008–.009	0.0024	0.0014–0.0021
Ag$^+$-Na$^+$	0.62	1.542	0.082	-0.05	0.083	0.09	0.0005	—

$V_0 = 15$ cm^3; $R_0 = (n - 1)V_0 = 7.7$ cm^3; all ΔV and ΔR in cm^3.

2.4.2 Index Change Profiles

Ion exchange produces waveguides with graded index boundaries, and various waveguide properties depend on the actual shape of the index profile. Although a detailed description of the exchange process in mathematical terms is reserved for a later chapter of this book, some considerations will be given here to particularities of the exchange from melts.

2.4.2.1 Effect of Relative Ion Mobilities

The change in concentration c of a dopant ion (with the maximum concentration set to unity) resulting from ion exchange obeys the following equation:

$$\frac{\partial c}{\partial t} = \frac{D_B}{1 - \alpha c}\left[\nabla^2 c + \frac{\alpha(\nabla c)^2}{1 - \alpha c} - e\frac{\vec{E}_{ext}}{kT}\nabla c\right] \qquad (2.11)$$

where D_B is a self-diffusion coefficient for the incoming ions, e is the electric charge of an electron, k is Boltzmann's constant, T is the absolute temperature, E_{ext} is an externally imposed electric field, and α is related to the ratio of the mobilities of the exchanging ions by

$$\alpha = 1 - \frac{D_B}{D_A} \qquad (2.12)$$

The only unknown parameters in equation (2.11) are the self-diffusion coefficients, D_i. They may be determined by measuring the depth and profile of planar waveguides made with dilute melts. By using dilute melts for this measurement, ordinary diffusion theory may be used, and true self-diffusion coefficients are obtained. The methods for characterizing planar guides are presented in Chapter 5.

The point we wish to make here is that eq. (2.11) is nonlinear in c, and the behavior of its solutions depends critically on the parameters. When there is no external field and α tends to 0, for instance, we come back to an ordinary diffusion equation with a constant diffusion coefficient.

The influence of α is best seen for 1-D exchange when no external field is present [14]. Representative results are shown in Figure 2.8. As α increases from 0, the shape of the solution goes from an erfc (1 minus the error function) profile

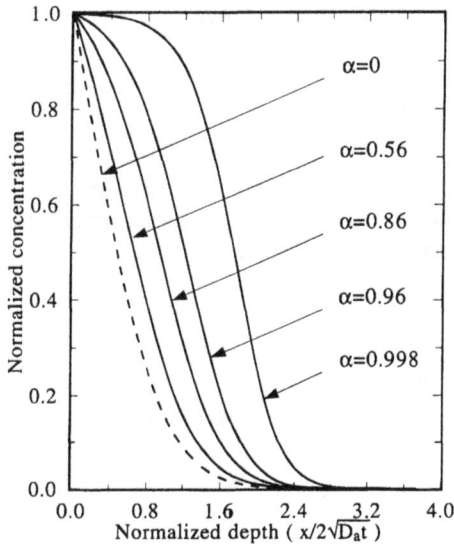

Figure 2.8 Influence of parameter α on 1-D profile shape [14].

toward a step function. This explains why different ion-exchange processes yield such different refractive index profiles. The value of α depends on the ions used *and* the substrate glass composition. In soda lime glasses for instance, undiluted silver-sodium exchange yields a parabolic profile ($\alpha \approx 0.5$) whereas potassium-sodium tends toward a Gaussian shape ($\alpha \approx 0.9$).

An externally imposed electric field also has an interesting effect in modifying the ion-exchange behavior. The concentration of dopant ions tends to be constant at its maximum value for a certain depth and decrease gradually thereafter. The relative sizes of the "flat" and graded regions depend on the relative strength of the two driving forces (thermal and electrical). As a rule of thumb, the depth of the flat, constant concentration region increases linearly with the product "voltage times duration," whereas the size of the graded region bounding it remains relatively unchanged (see Fig. 2.9). Therefore, it is worthwhile to investigate the purely thermal exchange separately, as in the following section.

2.4.2.2 Effect of Temperature and Duration

In many cases, it is unnecessary to undertake a full analysis with equation (2.11) to characterize ion exchanged waveguides. Once a reasonable value of α is available, an approximating function can be determined for the index profile. As an example, we use potassium-sodium exchange in an ordinary (Fisher brand) soda lime microscope slide [8]. For this case, α is close to 0.9 and a Gaussian function gives a good fit to the resulting index profile in one dimension. The profile is defined as follows:

$$n(x) = n_s + \Delta n \exp - \frac{x^2}{d^2} \qquad (2.13)$$

with x increasing from zero at the surface of the substrate, n_s the substrate index, Δn is the maximum index change, and d is the effective depth of the waveguide. It is found that d obeys the following rule:

$$d = \sqrt{D_e t} \qquad (2.14)$$

with D_e an "effective diffusion coefficient." Additionally, because the process is diffusion driven, this coefficient has an Arrhenius-type temperature dependence:

$$D_e = C_1 \exp - \frac{C_2}{T} \qquad (2.15)$$

where C_2 is proportional to the activation energy of the process. Figures 2.10 and

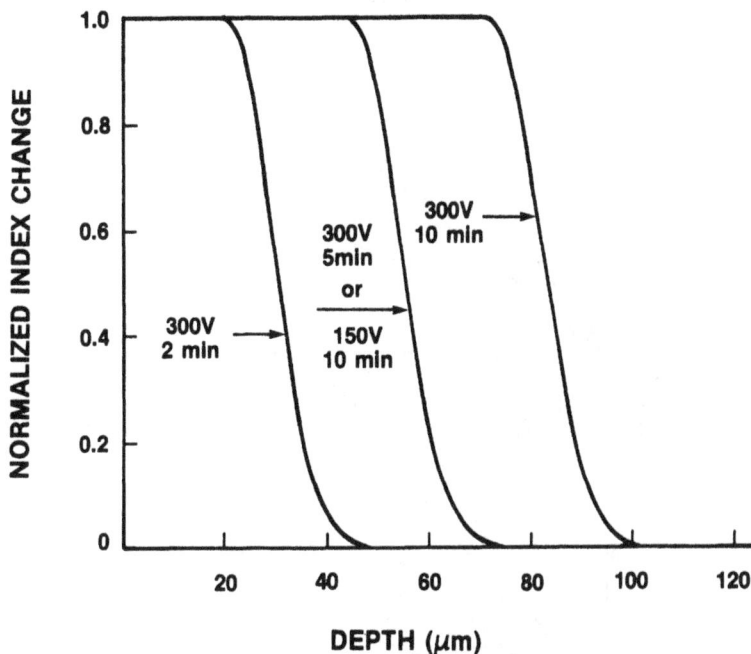

Figure 2.9 Depth profile for field-assisted exchange (silver-sodium, exchange conditions described in [10]).

2.11 show how the measured values of d and D_e follow equations (2.14) and (2.15) respectively. Practically, D_e is calculated from the slopes in Figure 2.10.

With these simple relationships, given a pair of values for T and t and the maximum index change Δn, a reasonable approximation for the profile may be written by using (2.13). It is then a simple matter to extract the mode properties of the waveguides by solving the wave equation for the given profile. Figure 2.12 shows a plot of measured effective indices for a set of waveguides, along with predicted values (obtained with a WKB analysis) from this approximate model. For shallow waveguides supporting few modes, the exact shape of the refractive index profile is less critical in determining modal properties than the values of d and Δn. It is why the agreement shown in Figure 2.12 is so good, almost exact in the limits of the measurement accuracy (10^{-4}).

In channel waveguides, it is more difficult to come up with such a simple, widely applicable approximate method of analysis. Generally we must use numerical modeling (for both the exchange process and the wave equation solution), and it is not much more complicated to solve the complete model, as represented by

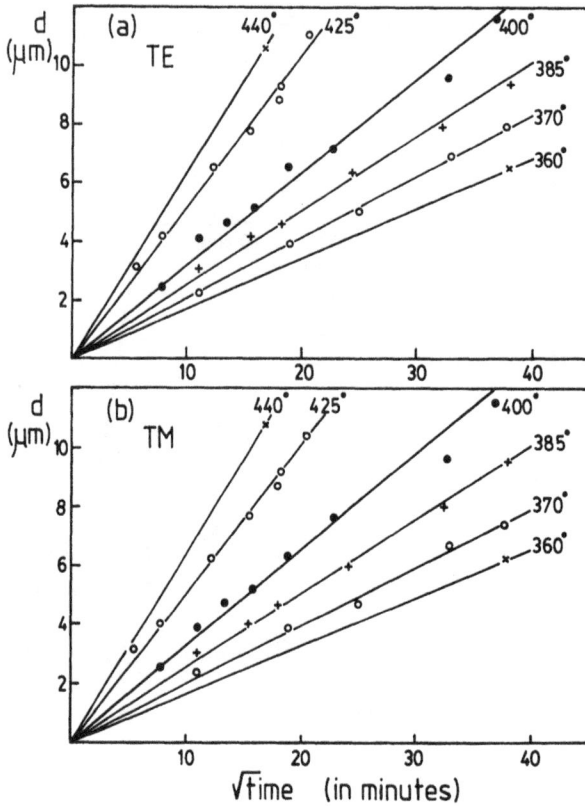

Figure 2.10 Experimental results showing the linear relationship between depth and the square root of exchange time for potassium-sodium exchange in microscope slides [8].

eq. (2.11), than some approximate one [6, 10]. In many cases, however, it is possible to use D_e defined previously as an approximation for D_a if the required accuracy does not exceed ± 10–20%.

2.4.3 Numerical Data for Waveguide Fabrication

A few ion exchange processes have been characterized in great detail in the literature. For these cases, Table 2.5 lists the parameters necessary to fabricate waveguides with various properties.

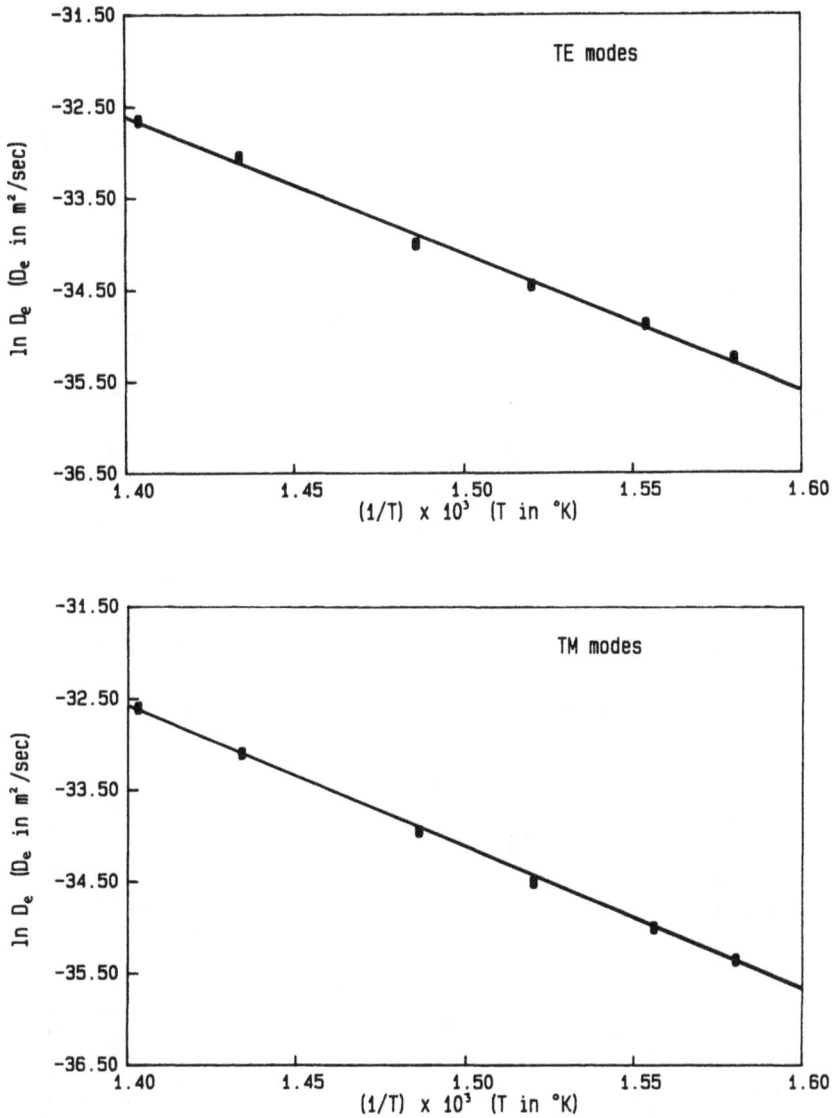

Figure 2.11 Temperature dependence of the effective diffusion coefficient (same case as Fig. 2.10) [8].

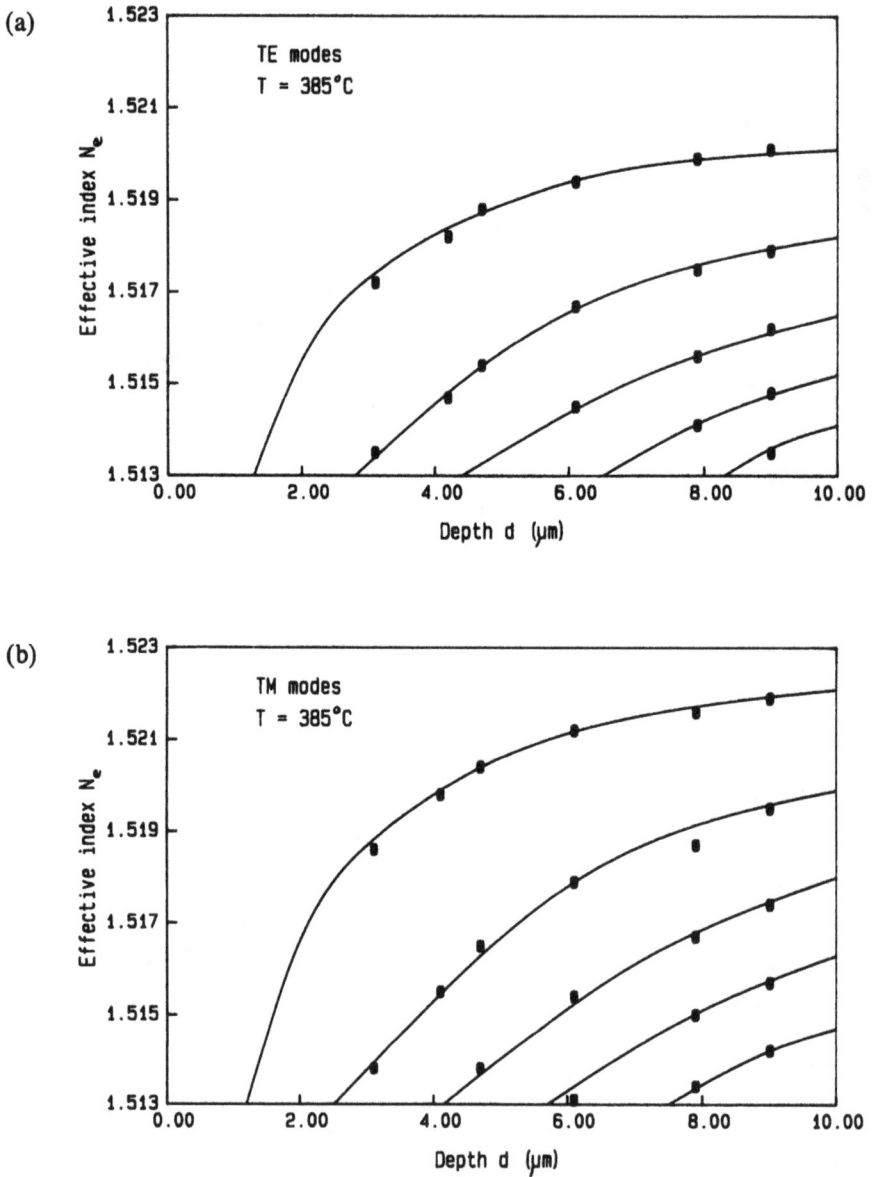

Figure 2.12 Measured effective indices (dots) and calculated values obtained with the Gaussian model for the index profile and a WKB analysis of the waveguide [15].

Table 2.5

Waveguide Properties for Some Common Ion-Exchange Salts and Substrates in Melts

Experimental Conditions	Waveguide Parameters	Ref.
Microscope slides (Chance) $n(633 \text{ nm}) = 1.5126$ Pure silver nitrate melt	$\Delta n(x) = \Delta n[1 - x/d - b(x^2/d^2)]$ $b = 0.64$; $\Delta n = 0.093$ $d = \sqrt{D_e t}$ $D_e = 2.26 \times 10^{-6} \exp(-1.02 \times 10^4/T)$ m^2/sec	[16]
Microscope slides (Chance) $n(633 \text{ nm}) = 1.5126$ Silver nitrate–sodium nitrate melt	(a) Very dilute (10^{-3} weight %) $\Delta n(x) = \Delta n \text{ erfc}(x/d)$ $\Delta n = 0.019$ $d = \sqrt{D_e t}$ $D_e = 1.59 \times 10^{-7} \exp(-9.10 \times 10^4/RT)$ m^2/sec (b) Large silver proportion (0.01–100 weight %) $\Delta n(x) = \Delta n[1 - x/d' - b(x^2/d'^2)]$ $b = 0.64$; $\Delta n = [.001 - .093]$ $d' = d[.26 + .74(.093/\Delta n)]$ $d = \sqrt{D_e t}$; $D_e = 2.26 \times 10^{-6} \exp(-1.02 \times 10^4/T)$ m^2/sec	[17]
Microscope slides (Fisher) $n(633 \text{ nm}) = 1.5131$ Potassium nitrate melt	$\Delta n(x) = \Delta n \exp(-x^2/d^2)$ $d = \sqrt{D_e t}$; $D_e(\text{TE}) = 7.8 \times 10^{-6} \exp(-1.489 \times 10^4/T)$ m^2/sec $D_e(\text{TM}) = 1.604 \times 10^{-6} \exp(-1.54 \times 10^4/T)$ m^2/sec $\Delta n = 8.4$–11.3×10^{-3}	[8]
Microscope slides (Labmate) $n(633 \text{ nm}) = 1.5125$ Silver ions electrolytically released in sodium nitrate melt	$\Delta n(x) = \Delta n \text{ erfc}(x/d)$ $d = 2\sqrt{D_e t}$ $D_e = 1.19 \times 10^{-7} \exp(-8.933 \times 10^4/RT)$ m^2/sec $\Delta n \approx 26.5 C_0$; where C_0 is the mole fraction of Ag$^+$ relative to Na$^+$ in the melt	[18]
Pyrex glass $n(633 \text{ nm}) = 1.4711$ Potassium nitrate melt	$\Delta n(x) = \Delta n \text{ erfc}(x/d)$ $d = \sqrt{D_e t}$ $D_e = 6.12 \times 10^{-16}$ m^2/sec ($T = 385°C$)	[19]
BK7 glass $n(633 \text{ nm}) = 1.5151$ Potassium nitrate melt	$\Delta n(x) = \Delta n \text{ erfc}(x/d)$ $d = \sqrt{D_e t}$ $D_e = 1.42 \times 10^{-15}$ m^2/sec ($T = 385°C$)	[19]

Microscope slides (Fisher
  Premium)
$n(633 \text{ nm}) = 1.512$
Silver nitrate: Potassium nitrate:
  Sodium nitrate
0.05:0.5:0.5 mole %
	No profile given, most likely
	$\Delta n(x) = \Delta n \text{ erfc}(x/d)$
	$\Delta n = 0.09$
	$d = 2\sqrt{D_e t}$
	$D_e = 2.21 \times 10^8 \exp(-9.959 \times 10^3/T)$ μm^2/hr
		[20]

Microscope slides (Fisher
  Premium)
$n(633 \text{ nm}) = 1.512$
Silver nitrate: Potassium nitrate:
  Sodium nitrate
0.01:0.5:0.5 mole %
	Same as previously
	$\Delta n = 0.078$
	$D_e = 1.09 \times 10^8 \exp(-9.828 \times 10^3/T)$ μm^2/hr
		[20]

Table 2.5
Continued

Epxerimental Conditions	Waveguide Parameters	Ref.
Microscope slides (Fisher Premium) $n(633 \text{ nm}) = 1.512$ Potassium nitrate melt	$\Delta n(x) = \Delta n \text{ erfc } (x/d)$ $\Delta n = 0.0082$ $d = 2\sqrt{D_e t}$ $D_e = 0.03 \ \mu\text{m}^2/\text{min } (T = 400°\text{C})$	[21]
Microscope slides (Fisher Premium) $n(633 \text{ nm}) = 1.512$ Potassium nitrate: Sodium nitrate 100:7.5 weight %	$\Delta n(x) = \Delta n \text{ erfc } (x/d)$ $\Delta n = 0.0074$ $d = 2\sqrt{D_e t}$ $D_e = 0.018 \ \mu\text{m}^2/\text{min } (T = 400°\text{C})$	[21]
Microscope Slides (Fisher Premium) $n(633 \text{ nm}) = 1.512$ Potassium nitrate: Sodium nitrate 100:15 weight %	$\Delta n(x) = \Delta n \text{ erfc } (x/d)$ $\Delta n = 0.0061$ $d = 2\sqrt{D_e t}$ $D_e = 0.015 \ \mu\text{m}^2/\text{min } (T = 400°\text{C})$	[21]
Corning 0211 glass $n(633 \text{ nm}) = 1.522$ Silver nitrate melt	$\Delta n(x) = \Delta n \text{ erfc } (x/d)$ $\Delta n = 0.062$ $d = 2\sqrt{D_e t}$ $D_e = 0.028 \ \mu\text{m}^2/\text{min } (T = 300°\text{C})$	[22]
Corning 0211 glass $n(633 \text{ nm}) = 1.522$ Cesium nitrate melt	$\Delta n(x)$ almost steplike $\Delta n = 0.025$ $d = 2\sqrt{D_e t}$ $D_e = 0.001 \ \mu\text{m}^2/\text{min } (T = 540°\text{C})$	[22]
Corning 0211 glass $n(633 \text{ nm}) = 1.522$ Potassium nitrate melt	$\Delta n(x) = \Delta n \text{ erfc } (x/d)$ $\Delta n = 0.0052$ $d = 2\sqrt{D_e t}$ $D_e = 0.14 \ \mu\text{m}^2/\text{min } (T = 400°\text{C})$	[21]
Corning 0211 glass $n(633 \text{ nm}) = 1.522$ Potassium nitrate: Sodium nitrate 100:7.5 weight %	$\Delta n(x) = \Delta n \text{ erfc } (x/d)$ $\Delta n = 0.004$ $d = 2\sqrt{D_e t}$ $D_e = 0.075 \ \mu\text{m}^2/\text{min } (T = 400°\text{C})$	[21]
Corning 0211 glass $n(633 \text{ nm}) = 1.522$ Potassium nitrate: Sodium nitrate 100:15 weight %	$\Delta n(x) = \Delta n \text{ erfc } (x/d)$ $\Delta n = 0.0035$ $d = 2\sqrt{D_e t}$ $D_e = 0.07 \ \mu\text{m}^2/\text{min } (T = 400°\text{C})$	[21]
Corning 0211 glass $n(633 \text{ nm}) = 1.522$ Thallium nitrate-Potassium nitrate melt	$\Delta n(x) = $ unspecified (d defined at $\Delta n/e^2$) $\Delta n = 0.04–0.09$ $d = 2\sqrt{D_e t}$ $D_e = 0.02–0.03 \ \mu\text{m}^2/\text{min } (T = 480–490°\text{C})$	[23]

ACKNOWLEDGMENTS

I wish to thank the many collaborators with whom I had the chance to work on ion exchanged glass waveguides over the years, especially Professors Gar Lam Yip of McGill University, who introduced me to this topic, and John Lit and Iraj Najafi for many fruitful discussions and collaborations.

REFERENCES

1. Kistler, S.S., "Stresses in Glass Produced by Nonuniform Exchange of Monovalent Ions," *J. Am. Ceram. Soc.*, Vol. 45, 1962, pp. 59–68.
2. Ramaswamy, R.V., and R. Srivastava, "Ion-Exchanged Glass Optical Waveguides: A Review," *IEEE J. Lightwave Technol.*, Vol. LT-6, 1988, pp. 984–1002.
3. Izawa, T., and H. Nakagome, "Optical Waveguide Formed by Electrically Induced Migration of Ions in Glass Plates," *Appl. Phys. Lett.*, Vol. 21, 1972, pp. 584–586.
4. Giallorenzi, T.G., E.J. West, R. Kirk, R. Ginther, and R.A. Andrews, "Optical Waveguides Formed by Thermal Migration of Ions in Glass," *Appl. Opt.*, Vol. 12, 1973, pp. 1240–1245.
5. Ross, L., "Integrated Optical Components in Substrate Glasses," *Glastech. Ber.*, Vol. 62, 1989, pp. 285–297.
6. Albert, J., and G.L. Yip, "Wide Single-Mode Channels and Directional Coupler by Two-Step Ion-Exchange in Glass," *IEEE J. Lightwave Technol.*, Vol. LT-6, 1988, pp. 552–563.
7. Albert, J., and G.L. Yip, "Insertion Loss Reduction Between Single-Mode Fibers and Diffused Channel Waveguides," *Appl. Opt.*, Vol. 27, 1988, pp. 4837–4843.
8. Yip, G.L., and J. Albert, "Characterization of Planar Optical Waveguides by K^+-ion exchange in Glass," *Opt. Lett.*, Vol. 10, 1985, pp. 151–153.
9. Yip, G.L., Noutsios, P., and K. Kishioka, "Characteristics of Optical Waveguides Made by Electric-Field-Assisted K^+-Ion Exchange," *Opt. Lett.*, Vol. 15, 1990, pp. 789–791.
10. Albert, J., and J.W.Y. Lit, "Full Modelling of Field-Assisted Ion-Exchange for Graded-Index, Buried Channel Optical Waveguides," *Appl. Opt.*, Vol. 29, 1990, pp. 2798–2804.
11. Fantone, S.D., "Refractive Index and Spectral Models for Gradient-Index Materials," *Appl. Opt.*, Vol. 22, 1983, pp. 432–440.
12. Albert, J., and G.L. Yip, "Stress-Induced Index Change for K^+-Na^+ Ion-Exchange in Glass," *Electron. Lett.*, Vol. 23, 1987, pp. 737–738.
13. Bradenburg, A., "Stress in Ion-Exchanged Glass Waveguides," *IEEE J. Lightwave Technol.*, Vol. LT-4, 1986, pp. 1580–1593.
14. Albert, J., and G.L. Yip, "Refractive-Index Profiles of Planar Waveguides Made by Ion-Exchange in Glass," *Appl. Opt.*, Vol. 24, 1985, pp. 3692–3693.
15. Albert, J., "Characterizations and Design of Planar Optical Waveguides and Directional Couplers by Two-Step K^+-Na^+ Ion-Exchange in Glass," Ph.D. Thesis, McGill University, Montreal, 1988.
16. Stewart, G., C.A. Millar, P.J.R. Laybourn, C.D.W. Wilkinson, and R.M. De La Rue, "Planar Optical Waveguides Formed by Silver-Ion Migration in Glass," *IEEE J. Quantum Electron.*, Vol. QE-13, 1977, pp. 192–200.
17. Stewart, G., and P.J.R. Laybourn, "Fabrication of Ion-Exchanged Optical Waveguides from Dilute Silver Nitrate Melts," *IEEE J. Quantum Electron.*, Vol. QE-14, 1978, pp. 930–934.
18. Lagu, R.K., and R.V. Ramaswamy, "Process and Waveguide Parameter Relations for the Design of Planar Silver Ion-Exchanged Glass Waveguides," *IEEE/OSA J. Lightwave Technol.*, Vol. LT-4, 1986, pp. 176–180.

19.   Gortych, J.E., and D.G. Hall, "Fabrication of Planar Optical Waveguides by K⁺-Ion Exchange in BK7 and Pyrex Glass," *IEEE J. Quantum Electron.*, Vol. QE-22, 1986, pp. 892–895.

20.   Jackel, J.L., "Glass Waveguides Made Using Low Melting Point Nitrate Mixtures," *Appl. Opt.*, Vol. 27, 1988, pp. 472–475.

21.   Najafi, S.I., "Optical Behavior of Potassium Ion-Exchanged Glass Waveguides," *Appl. Opt.*, Vol. 27, 1988, pp. 3728–3731.

22.   Li, M.J., W.J. Wang, and S.I. Najafi, "Ion-Exchanged Glass Waveguides with Spin-Coated Phosphate Overlays," presented at the IEEE International Workshop on Photonic Networks, Components, and Applications, Montebello (Quebec), Oct. 1990.

23.   Bourhis, J.-F., and S.I. Najafi, private communication, 1991.

Chapter 3
Silver-Film Ion-Exchange Technique
Seppo Honkanen

Nokia Research Center, Espoo, Finland

3.1 INTRODUCTION

Ion-exchanged glass waveguides typically are fabricated with molten salts as the ion sources. In the case of electric-field, assisted silver ion exchange silver thin films also are used extensively as ion sources. In principle, other metal films can be used as ion sources as well, and at least copper films have been applied in fabrication of ion-exchanged glass waveguides. The process characteristics with other metal film ion sources should be similar to silver, but some special arrangements, such as vacuum conditions, may be required in the fabrication process.

In this chapter we give an overview of the silver-film ion-exchange technique. First we give a brief introduction and a review of the research on silver-film ion-exchange technique. Section 3.2 presents the basics of the process, especially in comparison with the molten salt ion sources. Sections 3.3 and 3.4 discuss the fabrication and properties of slab waveguides and channel waveguides in more detail. Section 3.5 briefly discusses the waveguide losses, and finally we conclude with a look at some new possibilities for the silver-film ion-exchange technique.

This chapter concentrates on the experimental work on silver-film ion exchange. For comparison with the theory, the calculated curves (e.g., refractive index profiles) are taken directly from the literature without going into detail on the modeling procedures. The theoretical details and the modeling methods of different ion-exchange processes and ion-exchanged waveguides are presented in Chapter 4 of this book.

Optical waveguide fabrication by silver ion exchange using deposited silver thin film as the ion source was first reported by Chartier *et al.* [1]. After the first

---

report of this solid-state silver ion exchange several groups [2–4] utilized the method. Findakly and Garmire [2] reported a series of loss measurements of planar waveguides, and they demonstrated the strong wavelength dependence of the losses of silver ion-exchanged waveguides. The channel waveguide fabrication using patterned silver stripes and an aluminum anode was demonstrated by Viljanen and Leppihalme [3]. They fabricated fiber-compatible graded-index multimode waveguides with a process involving a postbake step to diffuse the exchanged ions deeper into the glass. The waveguides can also be patterned on deposited mask layers, such as aluminum. This has the advantage that the ion-exchange process can be performed in the vacuum chamber simultaneously with the silver-film evaporation step, as was demonstrated by Pitt, Stride, and Trigle [4]. Although the silver-film ion exchange has been regarded as a promising method for waveguide fabrication since it was first introduced, only a few, and mainly multimode, integrated optical components utilizing the method have been demonstrated. An example of the fabricated multimode components, a five wavelength demultiplexer [5], is sketched in Figure 3.1.

In recent years, the activity has increased in the field of silver-film ion exchange, because the technique offers unique possibilities to control the process and tailor the waveguide index profiles and geometries. The theoretical analysis of the process emphasizing the differences with molten salt ion sources has been given in [6] for slab waveguides and in [7] for channel waveguides. A lot of experimental work has been performed both for slab [8, 9] and channel [7, 10] waveguides. Also a new electrode structure, in which $NaNO_3$-melt is used as an anode, has been proposed [11]. This method enables the fabrication of buried channel waveguides with a nearly circular cross section in a one-step process [12–14]. Recently, the fabrication of single-mode components, that is, power splitters [15] and wavelength multiplexers [16], by the silver-film ion-exchange technique has also been demonstrated.

3.2 BASICS OF THE SILVER-FILM ION-EXCHANGE PROCESS

3.2.1 Comparison with Molten Salt Ion Sources

The fabrication of a planar optical waveguide into glass by the silver-film ion-exchange technique is sketched in Figure 3.2. To get a desired refractive index profile for the waveguide, the evolution of the Ag^+ concentration, which is almost linearly related to the refractive index profile, has to be known. The resulting Ag^+ concentration profile depends on the ion source used, because the nature of the source strongly affects the transition of ions into the glass. The basis of different ion-exchange processes is presented in Chapter 2 of this book. Here we will only

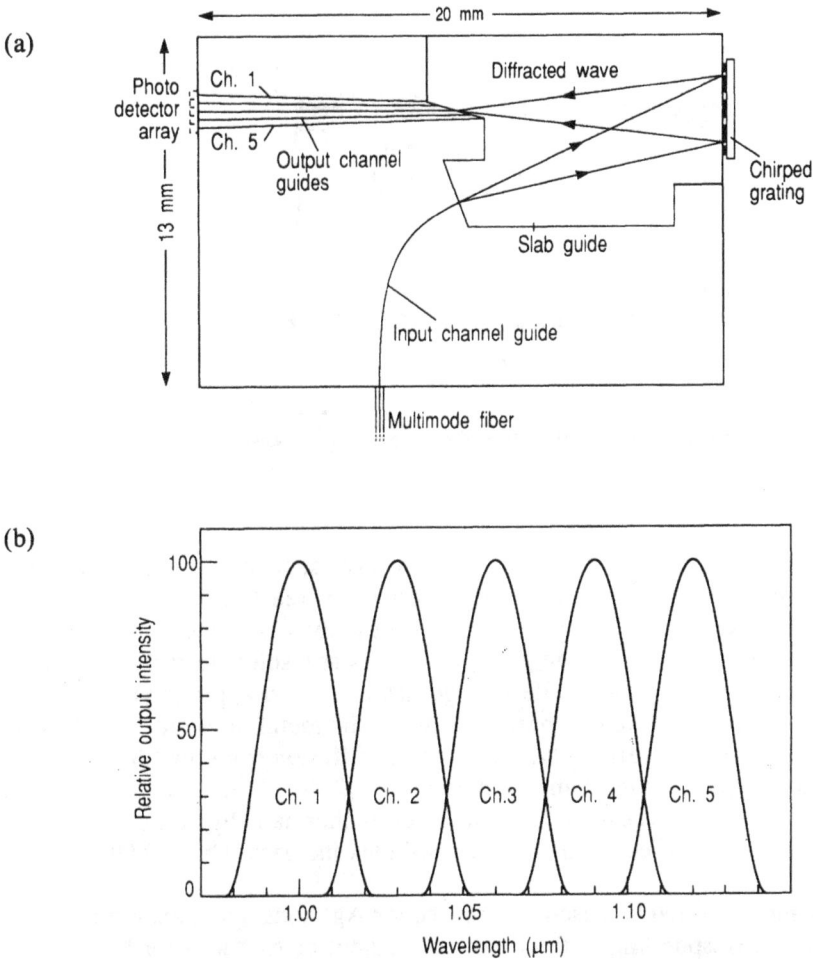

Figure 3.1 (a) Schematic of a multimode wavelength demultiplexer using a chirped grating and a silver-film ion-exchanged waveguide. (b) Measured characteristics of the fabricated device. (From [5].)

briefly summarize some special features of the silver-film ion-exchange technique, in comparison with the molten salt ion sources.

In silver ion exchange, a fundamental difference between a molten salt and a metallic film source is that the latter contains no silver ions and the exchange of ions between the source and the glass is not possible. When silver thin film is used

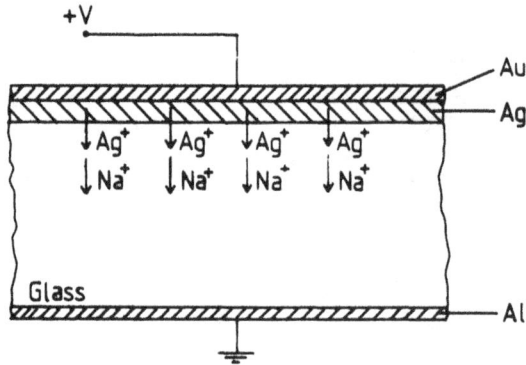

Figure 3.2 Schematic representation of the silver-film ion-exchange process.

as an ion source, the process needs electric-field assistance, and the silver ions are introduced into the glass only by an electrochemical reaction in which silver is oxidized: $Ag \rightarrow Ag^+ + e^-$. The process can be described as an electrolysis of Ag^+ ions from the silver anode into the glass, which acts as a solid electrolyte. To start the silver-film migration, the anode voltage has to exceed approximately 1 V to overcome the chemical potential barrier between the metal films and the glass. It has been experimentally verified that the thermal diffusion from the film into glass is negligible at typical ion-exchange temperatures [2, 8]. At higher temperatures (about 500°C) waveguides can be fabricated purely thermally by first oxidizing the silver film. With this process only a very low-index increase, about 0.001, can be obtained [2].

When molten $AgNO_3$ is used as a source for Ag^+ ions, the maximum surface concentration corresponding to the total replacement of Ag^+ ions for Na^+ ions is reached very rapidly due to the diffusion. In the case of silver-film ion exchange, there is no purely thermal diffusion from the source, and the surface concentration is determined by the electric-field strength and the process duration. Even with relatively high electric fields it takes a finite time to reach the maximum surface concentration. Figure 3.3(a) shows the calculated surface concentration of a slab waveguide as a function of the ionic current density [17]. Other parameters in the calculation were process duration $t = 1000$ s, initial sodium concentration in glass $c_0 = 5.0 \times 10^{-3}$ mol/cm^3, self-diffusion coefficient of silver ions $D_{Ag} = 2.0 \times 10^{-3}$ μm^2/s, and ratio of diffusion coefficients of silver and sodium ions in glass $M = 0.7$. For qualitative comparison Figure 3.3(b) shows the measured refractive index increase at the surface as a function of the applied voltage [8]. The experiments were performed at 275°C, using standard soda lime glass as a substrate.

(a)

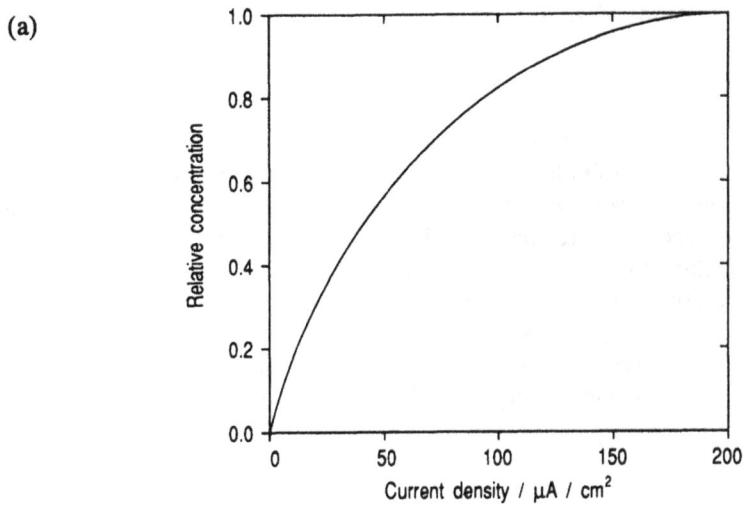

(b)

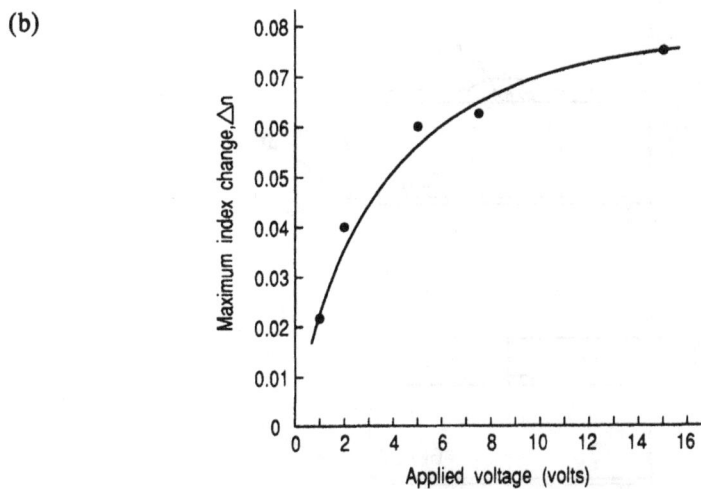

Figure 3.3 (a) Calculated surface concentration C_{Ag}/C_0 as a function of the ion current density. (From [17].) (b) Measured surface index change as a function of applied voltage. (From [8].)

3.2.2 Different Electrode Configurations

3.2.2.1 Dry Processes

One advantage of the silver-film ion-exchange technique is that it can be a totally dry process without molten salt electrodes, which makes the process easy to accomplish and well suited for mass production. Unlike with molten salts no special sample holders are needed in the electric-field assisted ion exchange. Typical electrode structures in the channel waveguide fabrication are sketched in Figure 3.4. The waveguide structures are patterned directly to the silver film or additional mask layers are used, as in the case of molten salt ion sources. A metallic film (e.g., gold) is often deposited over the silver electrode to ensure an electrical contact to all parts of the slowly disappearing silver film and to prevent oxidation of the

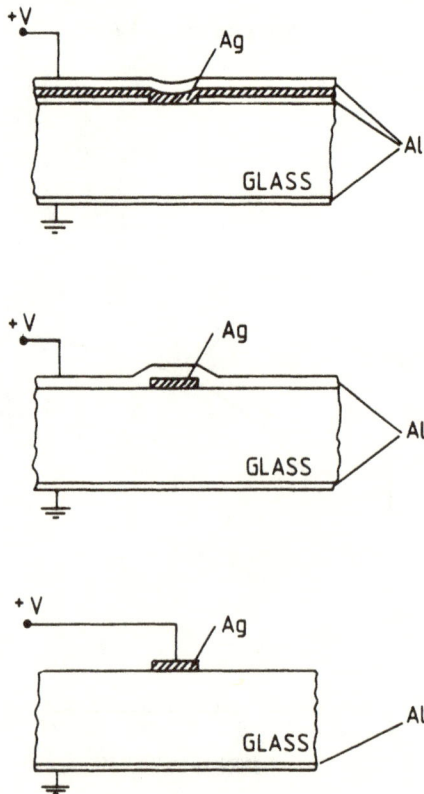

Figure 3.4 Typical electrode structures in dry silver-film ion exchange.

film. However, we have observed that, if the whole thickness of the silver film is not consumed, reproducible results also are obtained without a cover metal. Several metals have been used as a thin-film back electrode in the dry processes. The choice of a proper cathode material is important, as some metallic films (e.g., aluminum) tend to deteriorate due to an accumulation of sodium metal. We have found that silver also is highly suitable as a cathode, and at least 5 μm thick anode silver films can be migrated into the glass reproducibly.

When masks are used in silver-film ion exchange, there is more flexibility in choosing the mask material than in the case of molten salt ion sources. For instance chromium dissolves in most of the molten salts typically used, but it effectively prevents the silver-film migration under typical ion-exchange conditions. This presents an interesting possibility to directly utilize the integrated circuit mask plates as substrates in ion-exchange process [18]. These chromium mask plates are patterned with high accuracy and reliability by mask manufacturers, and the copies of the designed master mask are delivered at rather low cost.

In the masked silver-film ion exchange it is possible to fabricate channel waveguides in a vacuum chamber in connection with the silver-film deposition. This enables the reproducible migration of large amounts of silver ions (corresponding to silver film thicknessess up to 5 μm) into the glass by successive evaporation–ion-exchange steps. Figure 3.5 shows a schematic of the evaporation–ion-exchange apparatus of [19]. In [19] the process consisted of several evaporation and ion-exchange steps, and during each evaporation step, the voltage was dis-

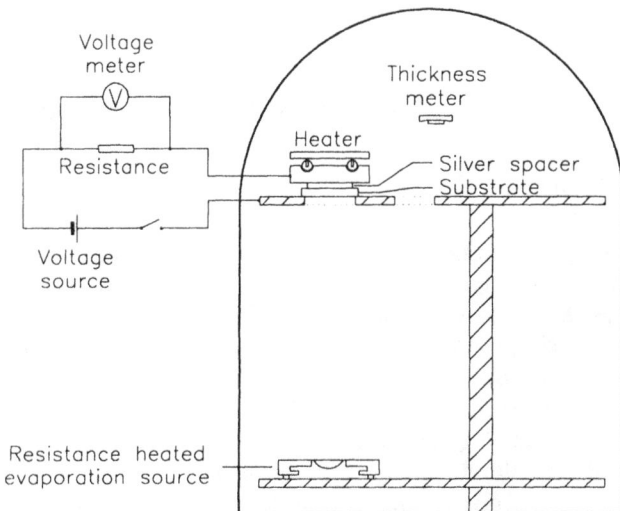

Figure 3.5 The experimental arrangement for silver evaporation and migration in vacuum. (From [19].)

connected from the electrodes. The deposited silver-film thicknesses varied between 200 nm and 500 nm, and the voltage used was between 50 and 200 V. The substrate glass was 0.6 mm thick Corning 0211. Figure 3.6 shows the electrical current measured during first three successive evaporation and ion-exchange steps. The exponential drop of the current is explained by the depletion current under the mask.

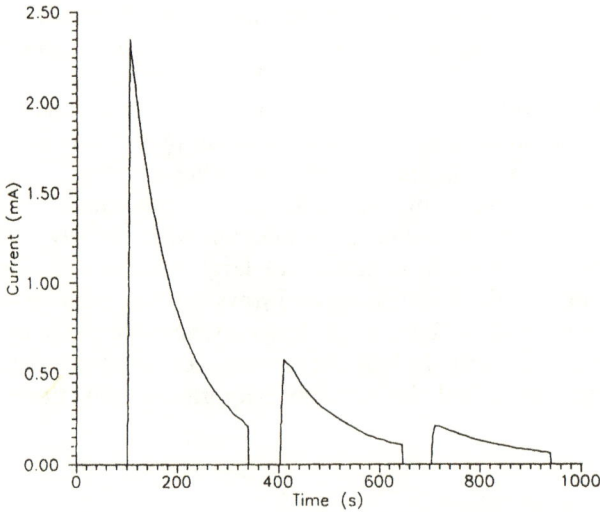

Figure 3.6 The electrical current during successive evaporation and ion-exchange steps. (From [19].)

3.2.2.2 Molten Salts as Electrodes

Figure 3.7 shows the experimental set-ups using molten salt electrodes in connection with silver-film ion exchange. Molten salts (e.g., $NaNO_3$) have been used as anodes [11] or cathodes [8], or as both electrodes [12–14]. The reason for using molten salt as a cathode is simply to prevent the deterioration of the cathode, which may occur if a metallic cathode is used. The use of $NaNO_3$ as an anode on top of a silver film presents interesting possibilities. The waveguides can be buried in a one-step ion-exchange process. The electric-field assisted burial starts immediately when the silver film is completely driven into the glass. Also, in channel waveguide fabrication this process gives more symmetric waveguide cross sections, because the shape of the electric-field distribution inside glass is more favorable with this electrode configuration. The side migration of silver ions is effectively eliminated.

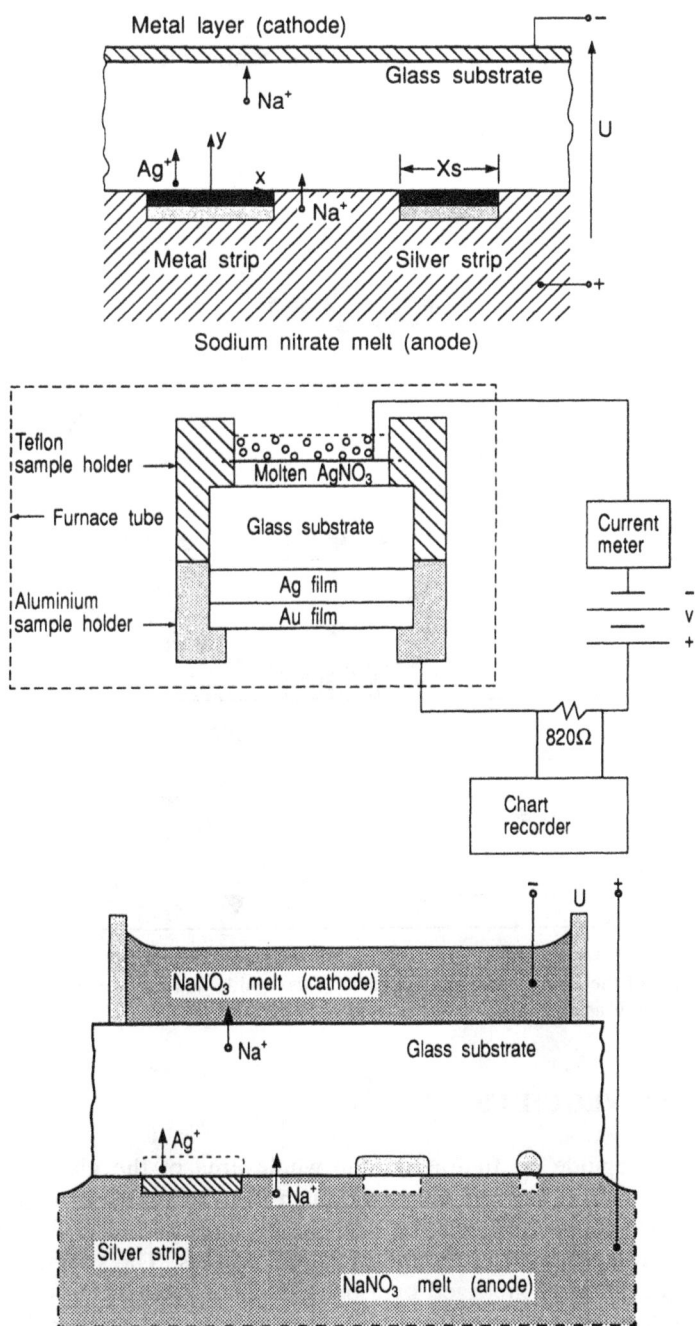

Figure 3.7 Possible electrode structures in silver-film ion exchange with molten salt electrodes. (From [11,8,12].)

The electric-field distributions in glass for this configuration and other silver-film configurations are illustrated in Figure 3.8.

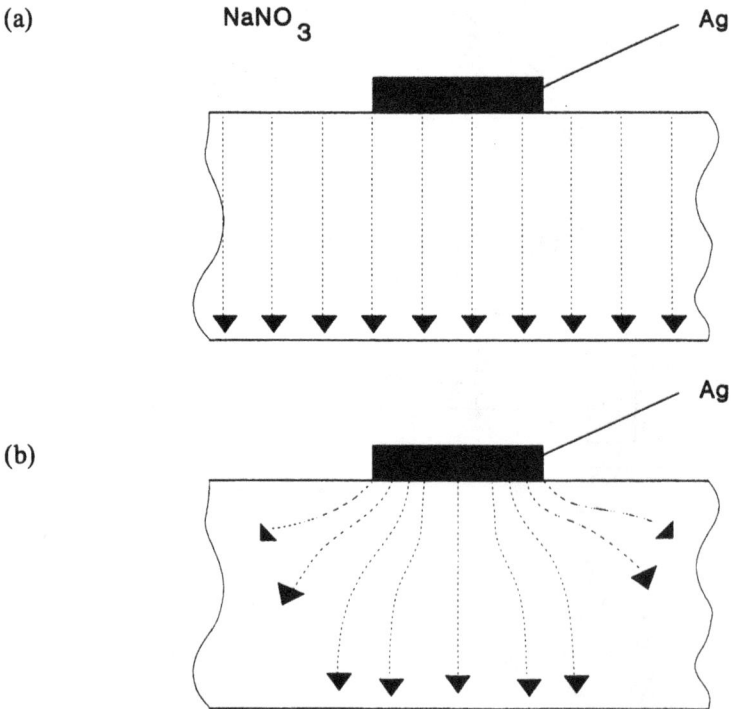

(a)

NaNO$_3$

Ag

(b)

Ag

Figure 3.8 Shape of the electric-field lines in channel waveguide fabrication (a) with and (b) without a salt melt as an anode.

3.3 SLAB WAVEGUIDES

In the slab waveguide configuration, the whole area of the glass surface is ion exchanged and acts as a waveguide. Although in most practical applications it is necessary to use channel waveguides, slab waveguides present useful information about the general behavior of the ion-exchange process. The slab waveguides can be easily and accurately analyzed by a prism-coupler method. In this section, we discuss the effects of the different process parameters of the silver-film technique on the refractive index profiles and the mode indices of the resulting slab waveguides. The parameters that may be varied in the process are its duration, temperature, and the ion current density.

3.3.1 Effect of the Process Duration

As the silver-film ion migration proceeds, the shape of the Ag^+ concentration profile in glass approaches a stationary state, provided that silver ions are less mobile in glass than sodium ions (which is usually the case). Figure 3.9 shows the calculated concentration profiles for different process durations [6]. The other parameters used in the calculations were $I = 150 \ \mu A/cm^2$, $D_{Ag} = 0.0026 \ \mu m^2$, and $M = 0.1$. The dashed lines in the figure are the profiles for stationary state and the solid lines represent numerically calculated concentration profiles. We see that the shape of the numerically calculated profile approaches the shape of the stationary state. However, it takes a considerably long time for the profile to reach the stationary state, and we prefer to use the more accurate numerically calculated profiles when designing the slab waveguides and in comparison of theoretical and

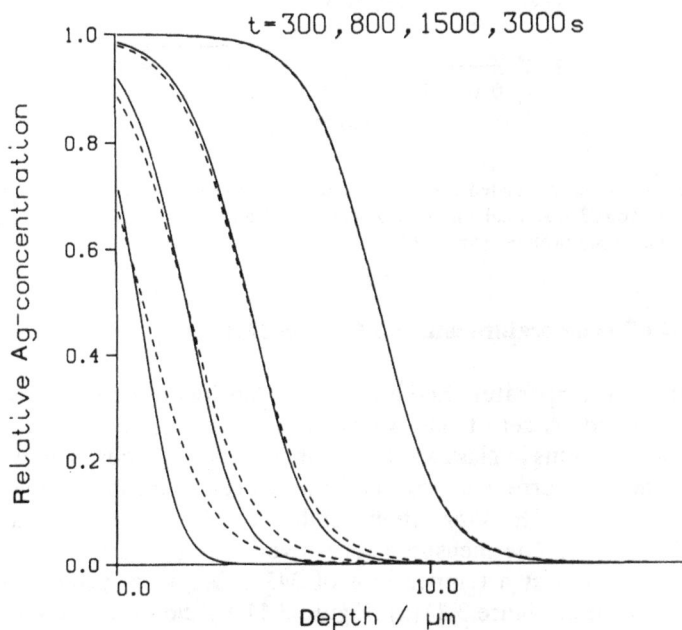

Figure 3.9 Numerically calculated concentration profiles for various ion exchange times (solid lines) compared with profiles with a stationary shape (dashed lines). (From [6].)

experimental refractive index profiles. Figure 3.10 shows the measured refractive index profiles for two slab waveguides fabricated into Corning 0211 glass [9]. The process durations were 1800 s and 3300 s and the other parameters were the same in both cases; that is, $T = 343°C$, $I = 114 \mu A/cm^2$. For comparison, the figure also shows the calculated profiles. In calculations, the material parameters were $M = 0.7$, $D_{Ag} = 0.0020 \mu m^2/s$, and $\Delta n_{max} = 0.05$.

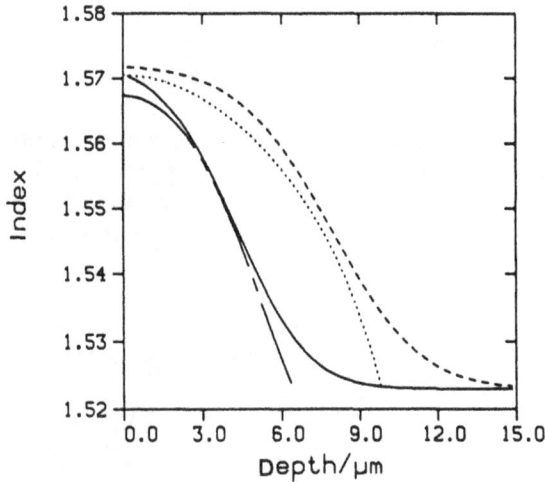

Figure 3.10 Comparison of calculated and experimental refractive index profiles for process durations 1800 s and 3300 s. Solid and dashed lines are calculated and chain-dashed and dotted lines are measured profiles. (From [9].)

3.3.2 Effect of Temperature and Ion Current Density

In [9] the effect of temperature and ion current density on the resulting waveguide profiles was studied. A set of slab waveguides was fabricated in such a way that the amount of silver ions in glass was the same in all waveguides. This was achieved by choosing the ion current density and the process duration to yield an equal product in each case. The fabrication conditions of these waveguides are summarized in Table 3.1. The measured refractive-index profiles of waveguides F7, F8, and F9 fabricated at a temperature of 343°C but with different ion current densities are shown in Figure 3.11(a). Figure 3.11(b) shows the measured profiles of waveguides F7, F10, and F11, which were fabricated with an ion current density of 114 $\mu A/cm^2$, but at different temperatures. The curves of Figure 3.11 clearly

Table 3.1
Fabrication Conditions of Waveguides F7–F11

Waveguide	I (μA/cm^2)	T(°C)	t(s)
F7	114	343	1800
F8	38	343	5400
F9	258	343	798
F10	114	314	1860
F11	114	367	1800

show how the refractive-index profile can be easily tailored with the process parameters without altering the amount of silver ions in glass. The calculated refractive-index profiles corresponding to the measured profiles of Figures 3.11(a and b) are presented in Figures 3.12(a and b).

In actual waveguide fabrication, it is often convenient to drive a desired amount of silver ions quickly into the glass and perform a subsequent purely thermal postbake to define the shape of the refractive index profile. The postbake diffuses the ions deeper into the glass and reduces the surface index for a more fiberlike index profile. To demonstrate the postbake process, the waveguides F7–F11 of Table 3.1 were held at 368°C for 70 min. The measured refractive-index profiles of these postbaked waveguides are shown in Figure 3.13. These profiles (and the amount of silver ions in glass) are nearly the same, which again confirms that no appreciable purely thermal diffusion of silver ions into the glass from the film has occurred during the field-assisted process step.

3.3.3 Profile Control by Ag Film Thickness

The total amount of silver ions in glass can be easily controlled by the Ag film thickness. When the whole film is driven into the glass and a constant anode potential is used, the resulting refractive-index profiles have a minor dependence on the temperature. This is because the diffusion coefficient of ions and the conductivity of glass have quite similar temperature dependencies. This behavior was demonstrated in [9] by driving 1 μm thick Ag films totally into the glass with a field of 76 V/mm at different temperatures. The processes were stopped when the current started to quickly fall off, indicating that the Ag film was almost totally consumed. Figure 3.14 shows refractive-index profiles of waveguides fabricated at temperatures 299°C (solid line) and 358°C (dashed line). The exchange times were 209 min and 35 min, respectively. Even these extreme differences in fabrication conditions result in only a slight variation in the refractive-index profiles.

(a)

(b)

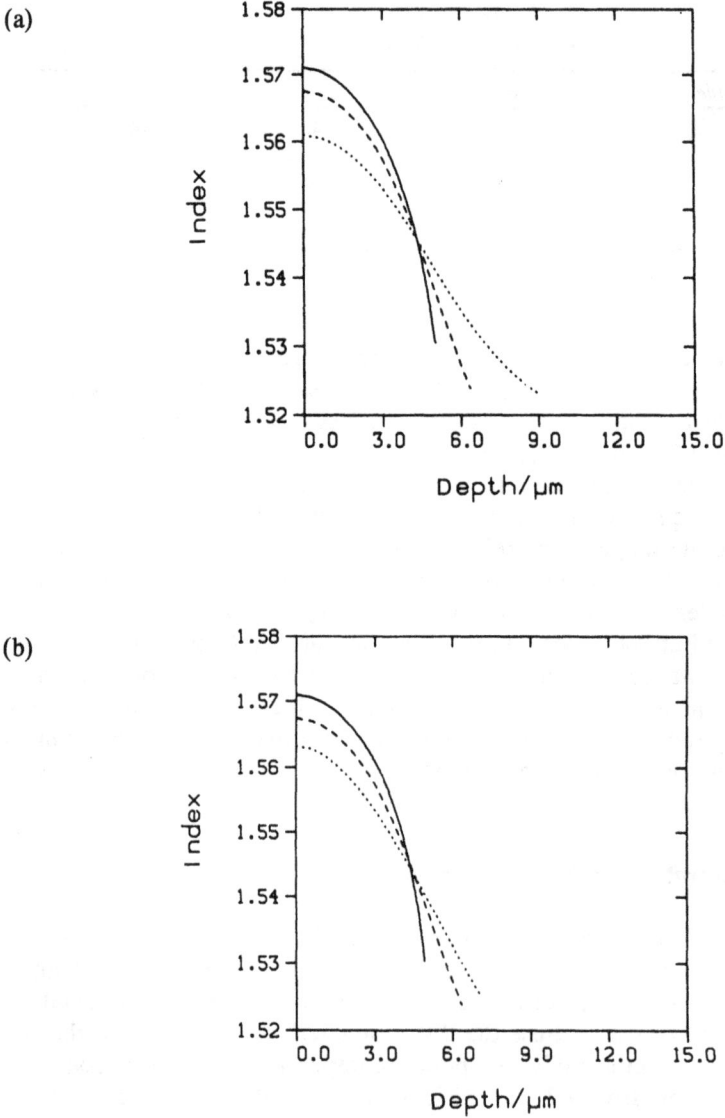

Figure 3.11 Measured refractive index profiles of waveguides: (a) F7 (dashed line), F8 (dotted line), and F9 (solid line); and (b) F7 (dashed line), F10 (solid line), and F11 (dotted line). (From [9].)

(a)

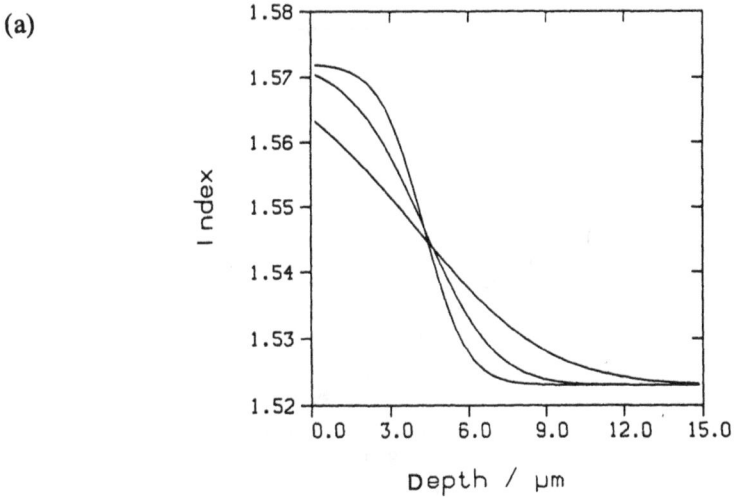

(b)

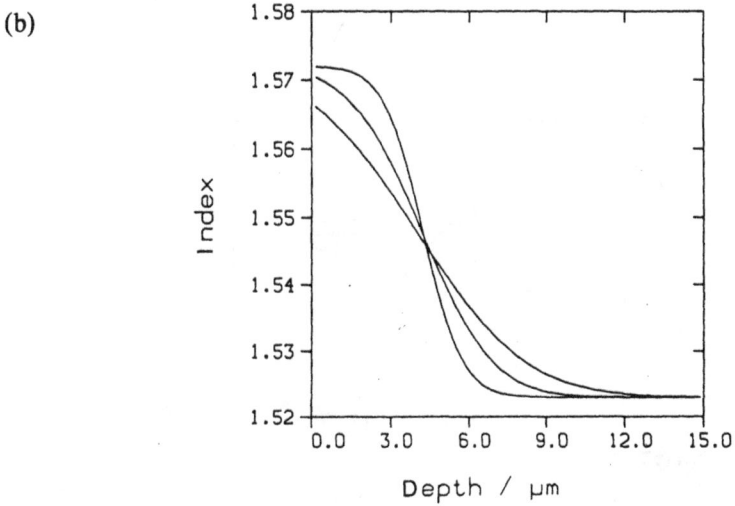

Figure 3.12 Calculated refractive index profiles corresponding (a) to Figure 3.11(a) and (b) to Figure 3.11(b). (From [9].)

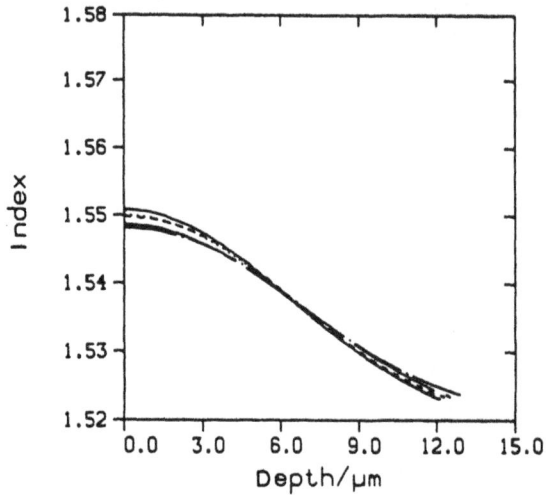

Figure 3.13 Measured refractive index profiles of waveguides F7 (dotted line), F8 (chain-dotted line), F9 (solid line), F10 (dashed line), and F11 (chain-dashed line) after a postbake of 70 min at 368°C. (From [9].)

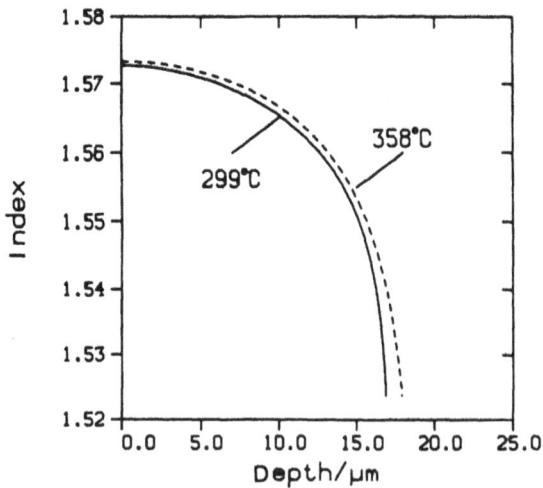

Figure 3.14 Measured refractive index profiles of two waveguides fabricated at different temperatures with 1 μm thick Ag film and an electric field of 76 V/mm. (From [9].)

3.3.4 Charge-Controlled Process

When silver film is used as an ion source, the amount of silver ions in glass can be accurately controlled by the electrical charge migrating into glass. The total charge can be integrated by measuring the current during the electric-field assisted process. Also, the field-assisted migration and a subsequent postbake step can be combined into a one-step process simply by disconnecting the anode voltage when the desired amount of silver ions is driven into glass. These characteristics are especially important in single-mode applications, because the fabrication of single-mode waveguides requires accurate control of the process. In [20] a set of slab waveguides was fabricated by this one-step charge-controlled process. In addition to charge integration, the diffusion (during the postbake step) was integrated as well by measuring the substrate temperature. This makes the process more insensitive to temperature changes in the furnace as the process goes on.

The process-controlling parameters were the exchanged charge and the effective postbake time. The effective postbake time was the postbake time for 343°C. The real postbake time was calculated by computer by integrating the diffusion utilizing the temperature dependence, $D_{Ag} = 40.6 \times 10^4 \ \mu m^2/s \times \exp(-1.18 \times 10^4 \ K/T)$ [9], of the diffusion coefficient in Corning 0211. The process parameters were chosen so that the waveguides were single moded or supported only a few modes at wavelengths 1.3 μm and 1.55 μm. The modal behavior of the fabricated waveguides was measured using a prism coupler set-up with a HeNe laser at wavelength 0.633 μm.

To compare the measured mode indices with theory, the curves for mode indices as a function of an effective postbake time for a given charge amount were calculated. The theoretical curves were obtained by calculating first the silver ion concentration profile with a one-dimensional version of the numerical model for ion-exchange processes [21] and using the material parameters obtained for silver ion exchange in Corning 0211 glass at the temperature of 343°C [9]. Then the concentration profile was transformed to a refractive-index profile, as in [22], and the mode indices were calculated numerically. As an example, Figure 3.15 presents the comparison for TE modes for charge amount of 90 mC/cm². Solid lines are the calculated curves, and circles and crosses represent the experimental mode indices obtained for temperatures 320°C and 250°C, respectively. The experimental results are in good agreement with the calculated curves. The figure also clearly shows the insensitivity of the process to the temperature used. The real postbake time in the experiment at 250°C was about ten times longer than at 320°C; for example, about 51 hrs at 250°C and about 4.7 hrs at 320°C corresponding to an effective postbake time of 120 min at 343°C. This significant difference in process temperature has only a slight effect on the waveguide mode indices.

The measured mode indices in Figure 3.15 are in good agreement with the theoretical curves, although a constant diffusion coefficient for silver ions was used

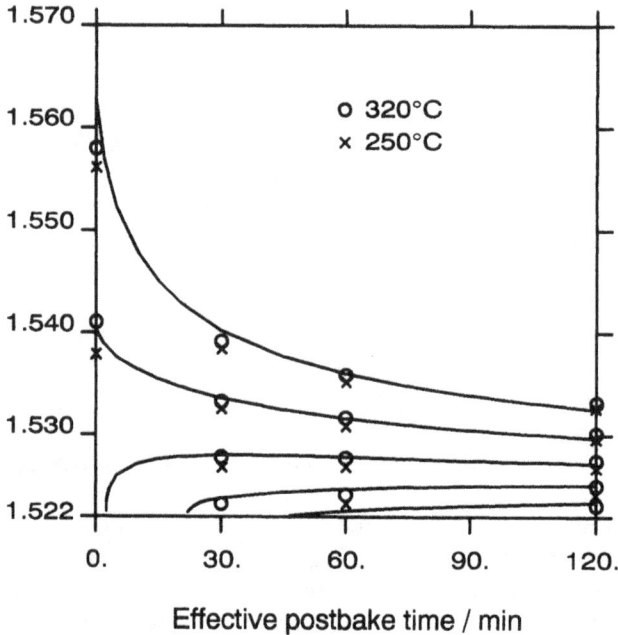

Figure 3.15 Mode indices of a slab waveguide (charge amount 90 mC/cm²) as a function of the effective postbake time. The solid lines are calculated curves and circles and crosses represent the measured indices. (From [20].)

in calculations. No evidence of reduced diffusion coefficient at low silver concentration is observed. To study this typical characteristic of silver ion exchange in more detail we fabricated a slab waveguide with charge amount of only 30 mC/cm³ and performed sequential postbakes at a temperature of 320°C. After each postbake the mode profile of the waveguide was measured at wavelength of 1.528 μm. In Figure 3.16 the measured mode profiles are compared with calculated mode profiles. We can see that after long postbake times the agreement between theory and experiments begins to fall off. Because the theoretical mode size is increasing more rapidly, the deviation can be explained by the reduced diffusion coefficient of silver ions at low Ag^+ concentrations.

3.4 CHANNEL WAVEGUIDES

3.4.1 Multimode Channel Waveguides

One characteristic of the Ag film ion sources is that they are rather limited and can be used completely in the ion-exchange process. This source-controlled nature

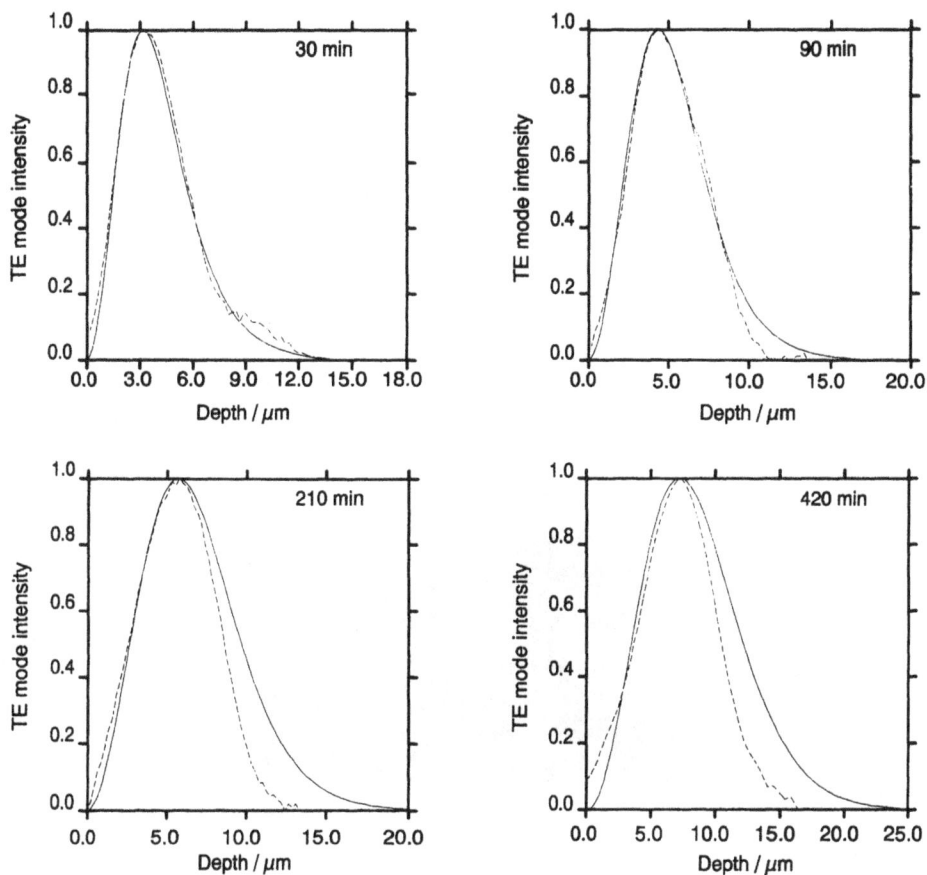

Figure 3.16 Comparison of measured (dotted lines) and calculated (solid lines) mode profiles of a slab waveguide with different postbake durations.

of the process makes it possible to control the waveguide depth profiles by the source dimensions [7]; that is, the Ag film thickness and the mask opening width. Exact control of the temperature and time is not necessary. Figure 3.17(a) shows the calculated film thickness profile and the advancement of the waveguide depth profile (step index approximation) as the process goes on. Owing to the stronger electric field near the edges of the source rather than in the middle of it, the Ag film is first exhausted from the edges. The penetration of the Ag^+ ion front is connected to the rate at which silver is consumed from the surface. The film thickness, the source width, and the glass substrate thickness used in the calculation were 0.3 μm, 30 μm, and 0.6 mm, respectively.

(a)

(b)

(c)

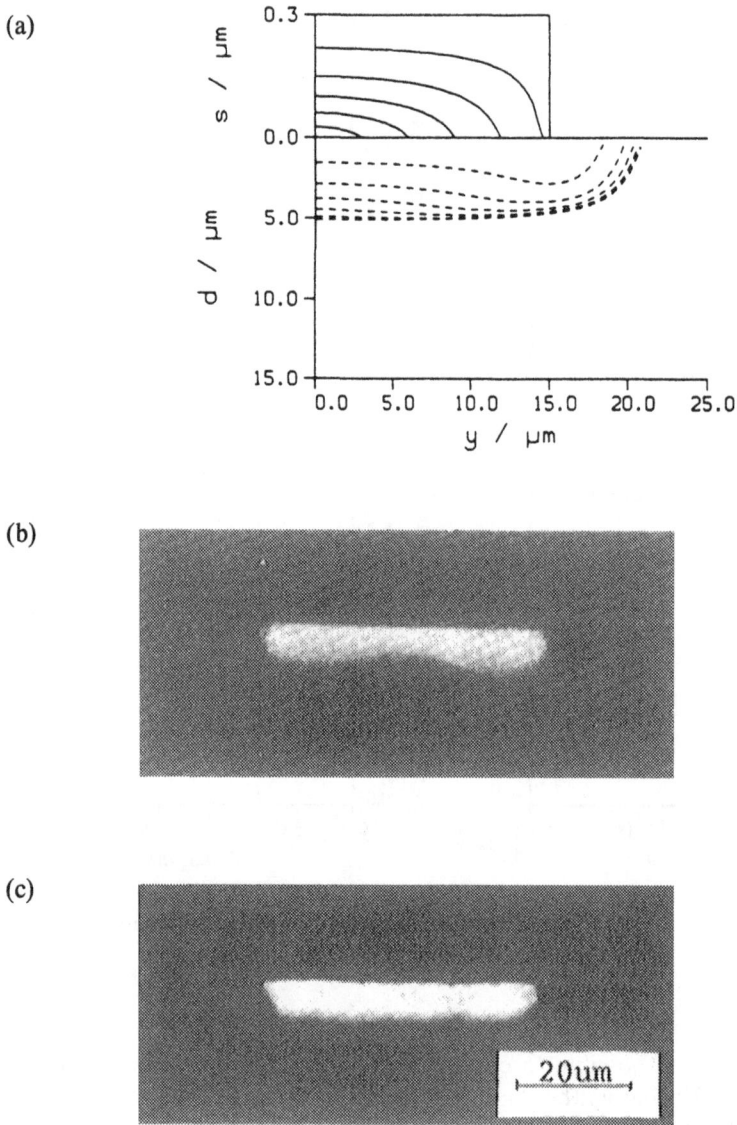

Figure 3.17 (a) Evolution of waveguide profile $d(y)$ and source film thickness $s(y)$ during channel waveguide fabrication (only one-half of the symmetrical cross section is shown). (b) Cross section of a waveguide fabricated with parameters similar to the calculation in Figure 3.17(a), but with the process not carried to completion. (c) Cross section of a fabricated waveguide when the whole thickness of the film was used. (From [7].)

Cross sections of channel waveguides fabricated with similar parameters into Corning 0211 glass are shown in Figures 3.17(b and c). In the waveguide of Figure 3.17(b) the process was not carried to completion and the cross section has the saddlelike shape typical of waveguides made by field-assisted ion exchange. This can be compared to the first depth profile in Figure 3.17(a). In Figure 3.17(c) the whole thickness of the silver film was used, and the waveguide cross section is quite uniform in depth, as in the calculation of Figure 3.17(a).

The source-controlled nature of the silver-film ion exchange offers unique possibilities to control the channel waveguide index profiles. This can be utilized when fabricating components having waveguides with different sizes into a single substrate, such as multimode star couplers and asymmetric power dividers. The planar star couplers are typically made by connecting a group of channel waveguides to both ends of a wide mixing region. The electric-field assisted ion exchange with molten salt ion sources is the most common fabrication method used. However, with molten salt ion sources the waveguides made through narrow openings become much deeper than the wider mixing waveguide; and the mixing waveguide is much deeper near its edges than in the middle of it. The differences in the waveguide depths cause additional losses and nonuniformities in the power distribution between different output ports. With silver-film ion sources the process can be source controlled, and the waveguides are more uniform in depth. The narrower waveguides become slightly shallower than the mixing waveguide, and the mixing waveguide is no deeper at its edges than in its middle.

Some multimode devices, like asymmetric optical couplers, require splitters having a large difference in cross-section areas of the branching waveguides. With molten salt ion sources in the electric-field assisted ion exchange this is difficult to accomplish. However, with the source-controlled silver-film ion exchange it is straightforward to fabricate waveguides of different sizes with nearly semicircular cross sections into the same substrate. This behavior is demonstrated in Figure 3.18, which shows the cross section of the waveguides fabricated into a single Corning 0211 glass through 5 μm wide and 30 μm wide mask openings and with 5 μm thickness of silver film [23].

For efficient coupling of the multimode waveguides to optical fibers, circular waveguide cross sections are needed. Circular waveguides may be obtained by joining two semicircular waveguides on top of each other. This was demonstrated in [3], in which the semicircular waveguides were fabricated by the silver-film ion-exchange method. There, the aim was to get channel waveguides compatible with graded index multimode fibers with 50 μm core diameter, and a postbake step was needed to reduce the maximum refractive index increase and get a proper index profile. The waveguides were fabricated into soda lime glass substrates. The electric-field assisted step was performed at 320°C with a voltage of 20 V for 30 min and with the middle electrode configuration of Figure 3.4. The postbake was per-

(a)

(b)

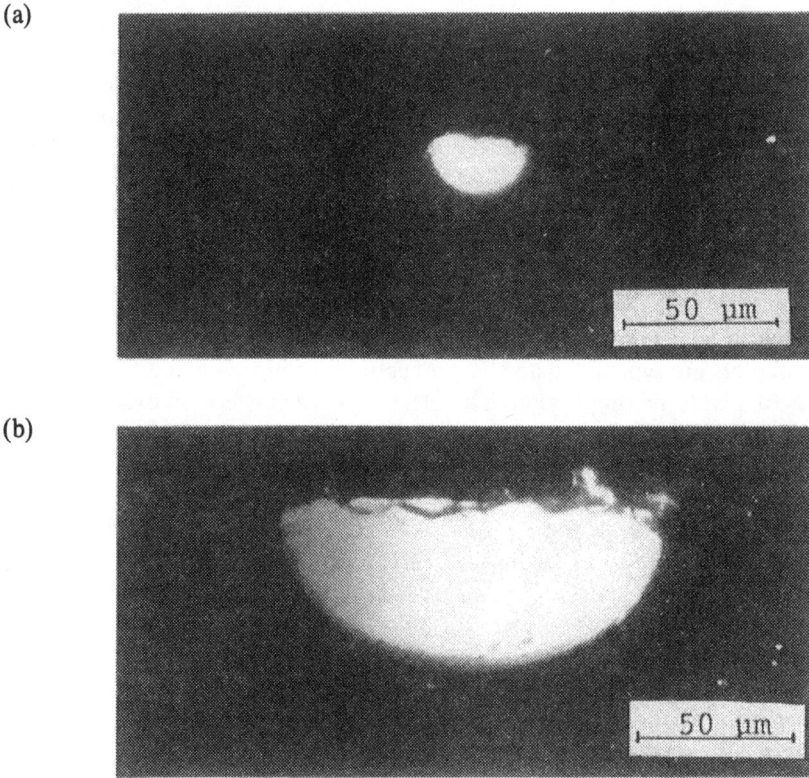

Figure 3.18 Cross section of waveguides fabricated into a same substrate through (a) 5 μm and (b) 30 μm wide mask openings and with 5 μm thickness of silver film. (From [23].)

formed at 360°C for 120 min. Figure 3.19 shows cross sections of circular waveguides formed by joining two substrates with semicircular waveguides on top of each other.

With the electrode configurations of Figure 3.7 (top and bottom) waveguides with circular cross sections buried below the glass surface can be obtained in a single process step. This is also an example in which the source-controlled nature of the silver-film ion-exchange technique can be attractively utilized. The cross section of a multimode waveguide [12], which was fabricated into a 0.5 mm thick borosilicate substrate glass (Desag, type 263), is shown in Figure 3.20. The process temperature, the used voltage, and the process duration were 347°C, 50 V and 15 min, respectively. The stripe width was 28 μm, and the film thickness was about 1 μm.

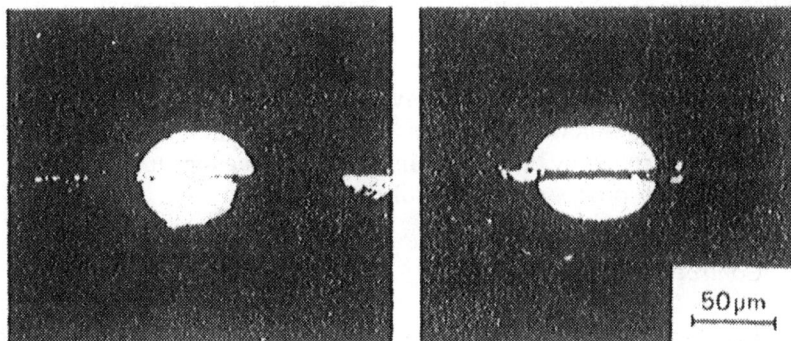

Figure 3.19 Circular waveguides formed by joining semicircular waveguides on top of each other. (From [3].)

Figure 3.20 Buried 35 μm wide circular waveguide made by one-step process. (From [12].)

3.4.2 Single-Mode Channel Waveguides

In fabrication of single-mode channel waveguides by ion-exchange techniques, two-step processes are typically needed. In the first step a desired amount of index-increasing ions is introduced into glass. The second step modifies the refractive-index profile to get channel waveguides that are compatible to optical fibers. Here we will discuss only the features that are special to the silver-film ion-exchange technique.

When single-mode waveguides are fabricated by conventional ion-exchange processes the process temperature has to be controlled very accurately, as even small variations in process temperature may change considerably the amount of dopant ions and the properties of the waveguides. The use of Ag-film ion sources is promising for single-mode applications, because the total amount of Ag^+ ions in glass can be controlled by the source dimensions or the ion current density used, and special one-step processes can be utilized.

3.4.2.1 Control by Source Dimensions

When source dimensions are used to control the process, narrow silver stripes are patterned on glass. The film thickness needed for single-mode waveguides is about 50–200 nm and the stripe width is around 5 μm. Using the electrode configurations of Figure 3.7 (top and bottom) it is possible to fabricate buried fiber compatible single-mode channel waveguides in a one-step process, which was demonstrated in [13]. Figure 3.21 shows the silver concentration profile measured by an electron microprobe of a buried channel waveguide. The waveguide was fabricated into a soda lime substrate glass using a 5 μm wide and 200 nm thick silver stripe. A mixture of 50 wt.% KNO_3 and 50 wt.% $NaNO_3$ was used as a cathode and pure $NaNO_3$ as an anode. The process temperature and the used electric field were 320°C and 30 V/mm, respectively. The measured near-field intensity distribution at 1.3 μm wavelength of the waveguide is shown in Figure 3.22, with comparison to optical fiber. The mode mismatch loss with the fiber is only about 0.1 dB.

3.4.2.2 Charge-Controlled Process

Fiber-compatible single-mode channel waveguides also can be fabricated by a one-step dry silver-film ion-exchange process, in which silver ions are diffused deeper into glass with a postbake immediately after the field-assisted migration by just disconnecting the voltage to the electrodes. Here the amount of silver ions in glass can be controlled very accurately by integrating electrical charge and measuring the current during the process. The ion source can be a patterned silver stripe or a mask. However, in channel waveguide fabrication it is difficult to directly measure the charge driven into glass from a narrow ion source. When silver stripes are used in configuration of Figure 3.4(bottom), the current is due mainly to the larger contact area needed on the substrate. This means that the process is not totally independent of the width of the stripe. Fortunately, the amount of silver ions in glass is quite insensitive to small changes in stripe width, because the electrical resistance of glass between a narrow ion source and a large area cathode has little dependence on small changes in the source width. This is illustrated in Figure 3.23, which shows the calculated amount of silver ions in glass as a function of stripe

(a)

(b)

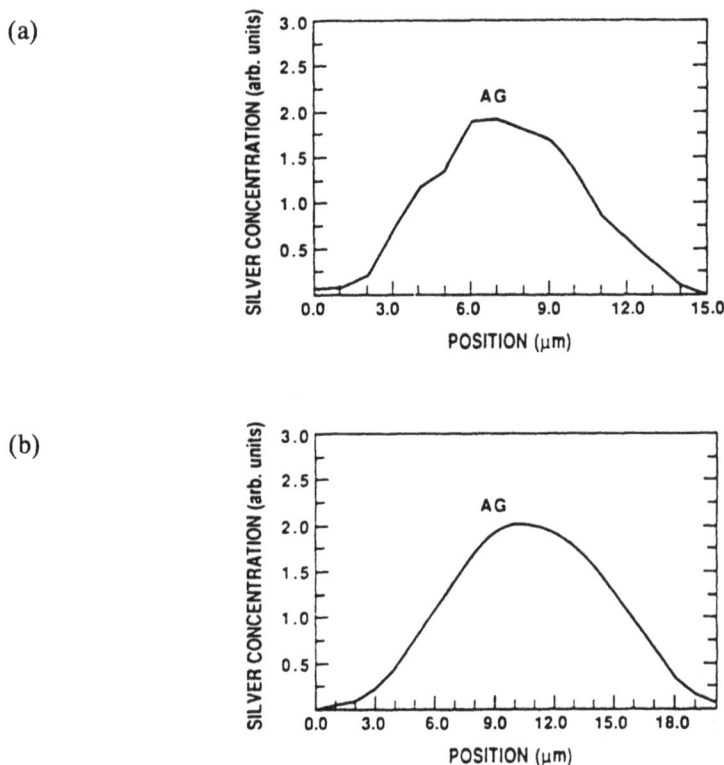

Figure 3.21 Silver concentration profile of a buried single-mode channel waveguide along (a) depth and (b) width directions. (From [13].)

width. The ion-exchange parameters used in the calculation were 5 V, 80 s, and 343°C, respectively, for voltage, duration, and temperature. The material parameters of Corning 0211 glass were used in the calculation. With a contact layer on the silver stripes or in a masked ion exchange, the depletion of sodium ions under the contact metal or mask gives an additional portion to the measured current. However, the lack of exact knowledge of the exchanged charge does not weaken the process's reproducibility as long as the same mask pattern is used. This will be the case in high-volume production.

In [15] the one-step dry ion-exchange process for single-mode channel waveguide fabrication was studied using Ti/W-alloy as a mask for ion exchange and using the same mask pattern throughout the experiments. The experiments were started using the duration of the electric-field assisted exchange as a process-controlling parameter and integrating the total charge by measuring the current during

(a)

(b)

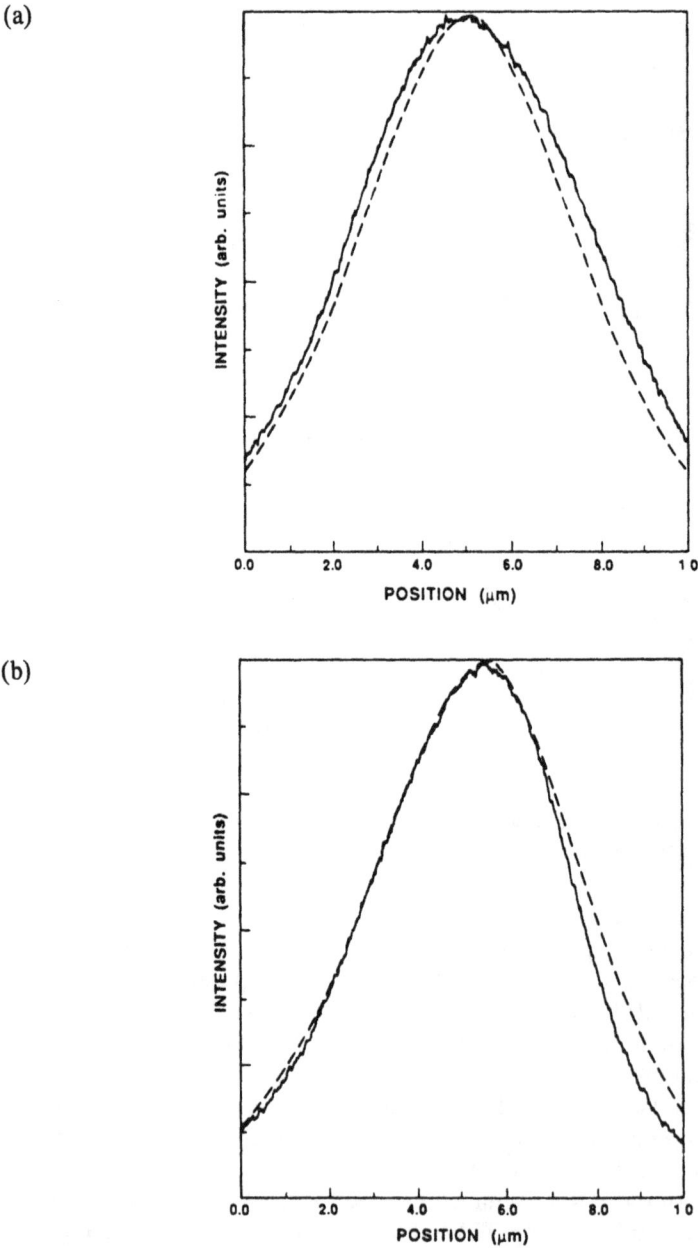

Figure 3.22 (a) Lateral and (b) vertical mode profiles of a buried waveguide (solid lines) in comparison with a single-mode fiber (dotted lines). (From [13].)

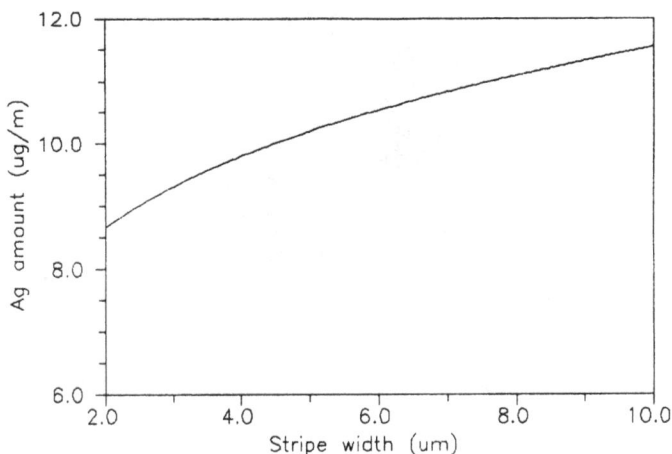

Figure 3.23 Calculated amount of silver ions in glass as a function of stripe width.

the process. After a few experiments the total charge could be used as a process-controlling parameter without knowing exactly the charge amount driven into glass through the waveguide mask opening. Figure 3.24 shows a measured mode profile at 1.528 μm wavelength of a channel waveguide, which was fabricated into 0.5 mm thick Corning 0211 glass substrate with a 4 μm mask opening, an anode voltage of 5 V (80 s), and a postbake duration of 6400 s at 343°C. With this mode profile the mode mismatch loss with standard telecom fiber is below 0.2 dB.

3.5 WAVEGUIDE LOSSES

A disadvantage of the silver ion-exchanged (e.g., silver-film ion-exchanged) wave-guides is the tendency of Ag$^+$ ions to neutralize to atomic silver during the ion-exchange process. The waveguide losses are increased due to the absorption of and scattering from colloidal silver particles. The main factors that influence the precipitation of silver ions are the substrate glass used, the process temperature, and the masking technique. The losses depend strongly on the wavelength used and have a maximum slightly above 400 nm, where the resonance wavelength for the precipitated small silver particles takes place. However, if substrate glass free of certain impurities is used and metallic masks are avoided, the loss increase due to the colloidal silver is negligible at wavelengths >800 nm. Using multimode silver-film ion-exchanged waveguides, losses below 0.1 dB/cm [2] for slab and 0.2 dB/cm [24] for channel waveguides have been measured. The strong wavelength dependence of the waveguide losses is illustrated in Figure 3.25, which presents a spectral

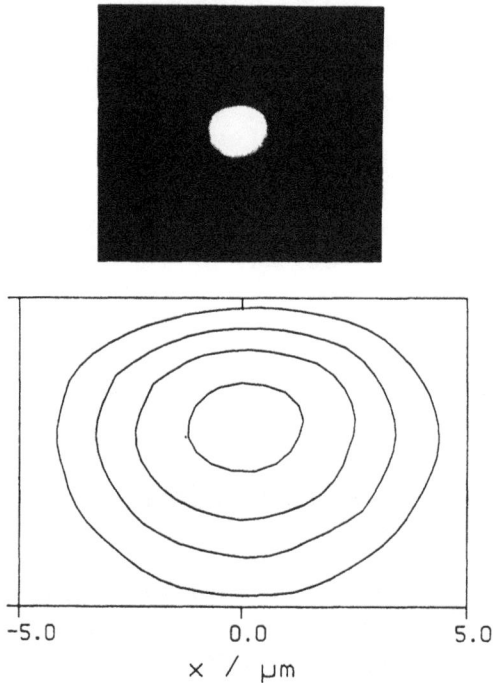

Figure 3.24 The near-field image and the measured mode intensity distribution of a channel waveguide fabricated by a one-step dry process. The contours are for normalized intensities 0.3, 0.5, 0.7, and 0.9. (From [15].)

loss measurement of a silver-film ion-exchanged multimode channel waveguide. The measured waveguide was fabricated into Corning 0211 glass with the electrode configuration of Figure 3.4 (bottom).

In single-mode channel waveguides fabricated by silver-film ion exchange the waveguide losses reported so far have been higher than 1 dB/cm. This is partly due to the fact that only a few studies on single-mode silver-film ion-exchanged waveguides have been reported. Also, when the ion source width is reduced by one order of magnitude compared to the multimode waveguides, the silver film quality (e.g., adhesion to glass) has to be much better. Very recently, the waveguide loss <0.2 dB/cm at 1.3 μm wavelength of a single-mode channel waveguide in Corning 0211 glass was measured [25]. The loss reduction was achieved by sputter cleaning the glass surface before the silver-film deposition and using dielectric mask in ion exchange.

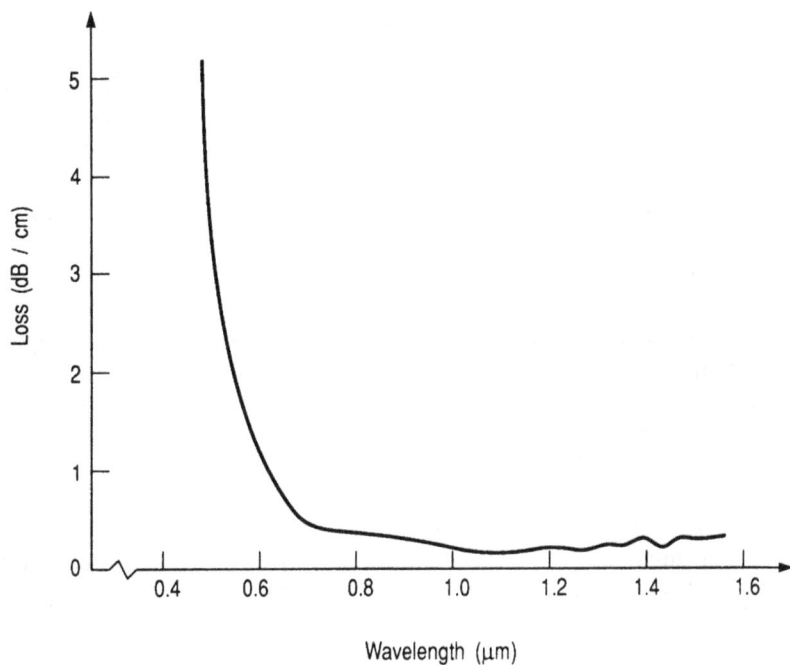

Figure 3.25 Spectral loss of a multimode silver-film ion-exchanged channel waveguide. (From [24].)

3.6 CONCLUSION

In recent years, substantial progress has been made in understanding the silver-film ion-exchange process. Good correlations have been obtained between the theoretical and experimental index profiles and waveguide properties. The silver-film ion exchange has shown some distinct advantages compared to conventional ion-exchange processes.

The silver ions penetrate into the glass only by electric-field assisted migration, and no appreciable purely thermal diffusion from the film into glass occurs. This makes it easy to control the total amount of silver ions in glass by the electrical current.

The silver-film ion exchange can be accomplished at a rather low temperature, it can be a totally dry process, and unlike molten salt processes, no special sample holders are needed during the electric-field assisted ion exchange. The process is well suited to high-volume mass production.

With silver film as the ion source in channel waveguide fabrication the choice of masking material is much more flexible than with molten salts. Recently, an interesting possibility for a masking technique in connection with silver-film ion exchange was proposed [26]. It was shown that standard photoresist can be used to mask the silver-film ion exchange by performing the process at quite a low temperature (about 140°C). The use of photoresist directly as a mask for the ion exchange makes the process as simple and accurate as possible, which is of crucial importance in passive glass-integrated optics. As a component example, a 1/8 single-mode power splitter with 4.1 dB facet-to-facet excess loss was reported.

When silver-film ion exchange is utilized in connection with rare-earth doped glasses, new possibilities arise. Most of the experiments with ion-exchanged rare-earth-doped glass waveguides have been performed with silicate glasses, although phosphate glasses can have much higher rare-earth concentrations. With phosphate glasses it is difficult to prevent the surface damage when conventional ion-exchange processes with molten salts as ion sources are used. The problem of glass surface deterioration is avoided by utilizing the silver-film ion-exchange technique. As an example, Figure 3.26 shows the mode-intensity distribution and spectral transmission curve of a single-mode channel waveguide, which was recently fabricated into an Er-doped phosphate glass [27]. These results are very encouraging for light amplification and laser applications at 1.5 μm wavelength region.

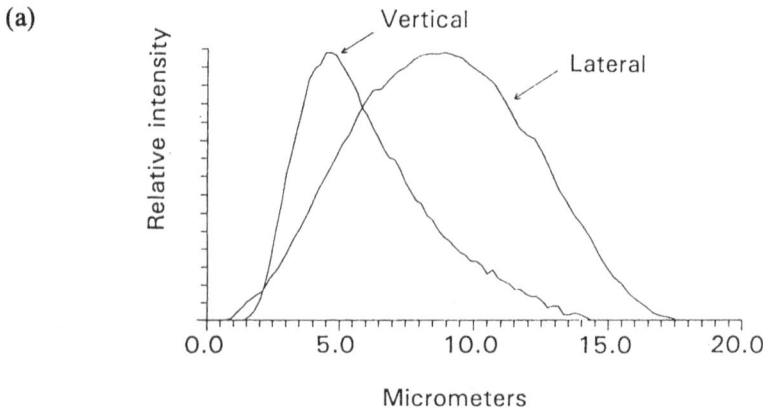

Figure 3.26 (a) Lateral and vertical mode profiles at 1.528 μm wavelength of an Er-doped single-mode waveguide. (b) The transmission spectrum of the waveguide around 0.98 μm and 1.55 μm wavelengths. (From [27].)

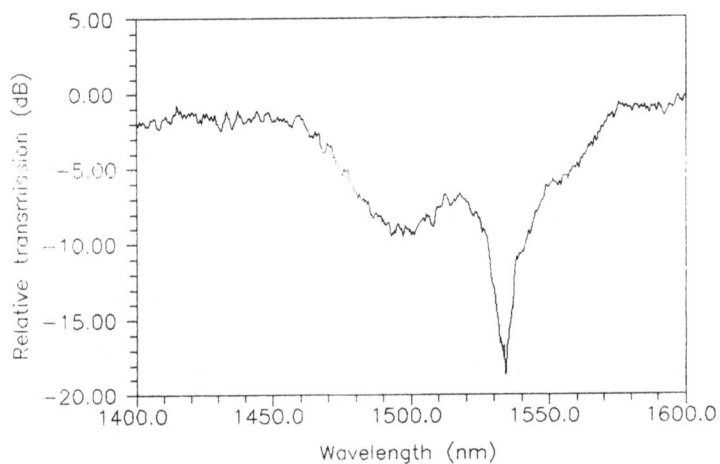

(b)

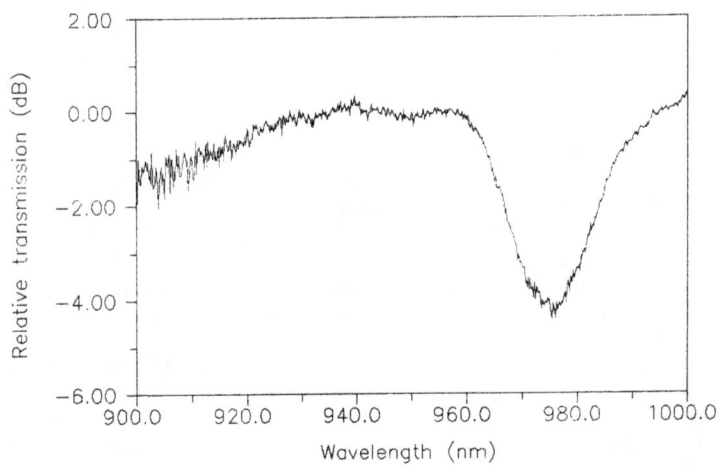

Figure 3.26 Continued.

REFERENCES

1. Chartier, G.H., P. Jaussaud, A.D. de Oliveira, and O. Parriaux, "Optical Waveguides Fabricated by Electric-Field Controlled Ion Exchange in Glass," *Electron. Lett.*, Vol. 14, 1978, pp. 132–134.

2. Findakly, T., and E. Garmire, "Reduction and Control of Optical Waveguide Losses in Glass," *Appl. Phys. Lett.*, Vol. 37, 1980, pp. 855–856.

3. Viljanen, J., and M. Leppihalme, "Fabrication of Optical Strip Waveguides with Nearly Circular Cross Section by Silver Ion Migration Technique," *J. Appl. Phys.*, Vol. 51, 1980, pp. 3563–3565.

4. Pitt, C.W., A.A. Stride, and R.I. Trigle, "Low Temperature Diffusion Process for Fabricating Optical Waveguides in Glass," *Electron. Lett.*, Vol. 16, 1980, pp. 701–702.

5. Suhara, T., J. Viljanen, and M. Leppihalme, "Integrated-Optic Wavelength Multi- and Demultiplexers Using a Chirped Grating and an Ion-Exchanged Waveguide," *Appl. Opt.*, Vol. 21, 1982, pp. 2195–2198.

6. Honkanen, S., A. Tervonen, H. von Bagh, and M. Leppihalme, "Ion Exchange Process for Fabrication of Waveguide Couplers for Fiber Optic Sensor Applications," *J. Appl. Phys.*, Vol. 61, 1987, pp. 52–56.

7. Tervonen, A., S. Honkanen, and M. Leppihalme, "Control of Ion-Exchanged Waveguide Profiles with Ag Thin-Film Sources," *J. Appl. Phys.*, Vol. 62, 1987, pp. 759–763.

8. Najafi, S.I., P.G. Suchoski, Jr., and R.V. Ramaswamy, "Silver Film-Diffused Glass Waveguides: Diffusion Process and Optical Properties," *IEEE J. Quantum Electron.*, Vol. QE-22, 1986, pp. 2213–2218.

9. Honkanen, S., and T. Tervonen, "Experimental Analysis of Ag^+-Na^+ Exchange in Glass with Ag Film Ion Sources for Planar Optical Waveguide Fabrication," *J. Appl. Phys.*, Vol. 63, 1988, pp. 634–639.

10. Forrest, K., S.J. Pagano, and W. Viehman, "Channel Waveguides in Glass via Silver-Sodium Field-Assisted Ion Exchange," *J. Lightwave Technol.* Vol. LT-4, 1986, pp. 140–150.

11. Pantchev, B., "Multimode Strip Waveguides Formed by Ion-Electro-Diffusion from Solid State Silver: Side Diffusion Reduction," *Opt. Comm.*, Vol. 60, 1986, pp. 373–375.

12. Pantchev, B., "One-Step Field-Assisted Ion Exchange for Fabrication of Buried Multimode Optical Strip Waveguides," *Electron. Lett.*, Vol. 23, 1987, pp. 1188–1190.

13. Zhenguang, H., R. Shirastava, and R.V. Ramaswamy, "Single-Mode Buried Channel Waveguide by Single-Step Electromigration Technique Using Silver Film," *Appl. Phys. Lett.*, Vol. 53, 1988, pp. 1681–1683.

14. Gunther, C., and D. Jestel, "Buried Waveguides Produced by a One-Step Field-Assisted Ag^+ Ion-Exchange in Glass," *Proc. SPIE—Int. Soc. Opt. Eng.* (USA), Vol. 993, 1988, pp. 2–6.

15. Honkanen, S., P. Pöyhönen, M. Tahkokorpi, A. Tervonen, and S. Tammela, "Single-Mode Glass Waveguides and Components by One-Step Dry Ion Exchange Technique," Proc.: Eighth Annual European Fibre Optic Communication and Local Area Networks Exposition, EFOC 90, Munich, June 27–29, 1990, pp. 135–137.

16. Tervonen, A., P. Pöyhönen, S. Honkanen, and M. Tahkokorpi, "A Guided-Wave Mach-Zehnder Interferometer Structure for Wavelength Multiplexing," *IEEE Photon. Technol. Lett.*, Vol. 3, 1991, pp. 516–518.

17. Tervonen, A., S. Honkanen, and M. Leppihalme, "Ion-Exchange Processes in Glass for Fabrication of Waveguide Couplers," *Proc. SPIE—Int. Soc. Opt. Eng.* (USA), Vol. 862, 1988, pp. 32–39.

18. Honkanen, S., A. Tervonen, H. von Bagh, A. Salin, and M. Leppihalme, "Fabrication of Ion-Exchanged Channel Waveguides Directly into Integrated Circuit Mask Plates," *Appl. Phys. Lett.*, Vol. 51, 1987, pp. 296–298.

19. Tammela, S., H. Pohjonen, S. Honkanen, and A. Tervonen, "Fabrication of Large Multimode Glass Waveguide by Dry Silver Ion Exchange in Vacuum," *Proc. SPIE—Int. Soc. Opt. Eng.* (USA), Vol. 1583, 1992.

20. Honkanen, S., A. Tervonen, P. Pöyhönen, M. Tahkokorpi, A. Rico, and M. Siilin, "Development of Glass Waveguide Components for Passive Optical Networks," Proc. Ninth Annual European Fibre Optic Communications and Local Area Network Conference, EFOC 91, London, June 19–21, 1990, pp. 229–232.

21. Tervonen, A., "A General Model for Fabrication Processes of Channel Waveguides by Ion Exchange," *J. Appl. Phys.*, Vol. 67, 1990, pp. 2746–2752.

22. Tervonen, A., P. Pöyhönen, S. Honkanen, M. Tahkokorpi, and S. Tammela, "Examination of Two-Step Fabrication Methods for Single-Mode Fiber Compatible Ion-Exchanged Glass Waveguides," *Appl. Opt.*, Vol. 30, 1991, pp. 338–343.

23. Honkanen, S., A. Tervonen, J. Harju, and P. Karioja, "Fabrication of a Multimode Asymmetric Coupler into Glass by Solid-State Silver Exchange," Proc. Seventh Annual European Fibre Optic Communication and Local Area Network Conference, EFOC/LAN 89, Amsterdam, June 12–16, 1989, pp. 300–302.

24. Viljanen, J., and M. Leppihalme, "Analysis of Loss in Ion-Exchanged Glass Waveguides," *Proc. European Conf. Integrated Optics*, Institution of Electrical Engineers, London, 1981, p. 18.

25. Pöyhönen, P., submitted for publication.

26. Pöyhönen, P., S. Honkanen, A. Tervonen, M. Tahkokorpi, and J. Albert, "Planar 1/8 Splitter in Glass by Photoresist Masked Silver Film Ion Exchange," *Electron. Lett.*, Vol. 27, 1991, pp. 1319–1320.

27. Honkanen, S., S.I. Najafi, P. Pöyhönen, G. Orcel, W.J. Wang, and J. Crostowski, "Silver-Film Ion-Exchanged Single-Mode Waveguides in Er-Doped Phosphate Glass," *Electron. Lett.*, Vol. 27, 1991, pp. 2167–2168.

Chapter 4
Theoretical Analysis of Ion-Exchanged Glass Waveguides
Ari Tervonen

Nokia Research Center, Espoo, Finland

For fabrication process development, for optimization of waveguide characteristics, and for design of integrated optics components, modeling is required. In ion-exchanged waveguides, the nature of refractive-index profiles is determined by the fabrication process, and waveguide profiles are tailored by complicated multistep processes. Because of this, modeling the ion-exchange processes is important. For linking the ion-exchange processes and the optical properties of the waveguides, knowledge about the relationship between ion concentration and refractive-index distributions is used. With the known refractive-index profiles of the waveguides, the theory of optical propagation in waveguides can be applied for modeling the waveguide properties and designing integrated optics components.

The purpose of this chapter is to introduce the modeling methods to be used in connection with the ion-exchanged waveguides. The aim is to describe modeling tools at a general level, so that they can be applied to a wide range of ion-exchange processes and for different guided-wave structures. Examples are provided to give insight into typical use of theoretical analysis.

4.1 MODELING ION-EXCHANGE PROCESSES

This section concentrates on the modeling of ion-exchange processes to develop ion concentration profiles in the glass during the waveguide fabrication. First, the physics of transport of ions inside the glass during the binary ion exchange is

---

described, and a general partial differential equation for ion exchange is derived. Next, another important aspect is considered: the ion exchange reactions at the interface of the ion source and the glass. This gives the boundary conditions to be used in solving the ion-exchange equation. One-dimensional slab waveguides are used for the examination of solutions, with examples of typical fabrication processes. Last, two-dimensional profiles of channel waveguides are discussed, starting with and extending from the tools and principles used in slab waveguide process modeling.

4.1.1 Diffusion and Migration of Ions in the Glass

The structure of glass is a network of glass formers, the most common of which are SiO_2 and B_2O_3, modified by other components of the glass composition. The alkali metal oxides present in the glass are in the form of monovalent alkali ions associated with the nonbridging oxygen atoms of the glass structure. These ions have a relatively high mobility at elevated temperatures.

The transport of ions in glass is now examined, taking into account the contribution of the electric field. The equation for the flux of ions from diffusion is

$$\bar{j} = -D\nabla c \tag{4.1}$$

where D is the self-diffusion coefficient, here assumed to be independent of the ionic concentration c. The flux produced by the electric field $\bar{E}$ is

$$\bar{j} = Dcq\bar{E}/fkT \tag{4.2}$$

Here q is the electric charge of the ion, k is Boltzmann's constant, T is the absolute temperature, and the Nernst-Einstein relation for the ionic conductivity σ has been used:

$$D = \sigma kT/cq^2 \tag{4.3}$$

The correlation factor f has been included. The reason for this is the difference of ionic mobility in diffusion and in being driven by an electric field. This is usually explained by the different mechanisms for the movement of ions [1]. In diffusion there is a correlation between the direction of successive jumps of ions from one site in the glass structure to another. In electrical conduction this correlation does not exist.

In binary ion exchange there are two species of mobile ions in the glass. These ions are considered to be monovalent, to occupy similar sites in the glass network, and to diffuse by the same mechanism, so that they are fully exchangeable. The unequal mobilities of the ions result in the electric field generated by diffusion of the ions. Flux equations for these ionic species denoted by subscripts A and B are written, with the charge of proton e,

$$\bar{j}_A = -D_A \nabla c_A + c_A D_A e\bar{E}/fkT \tag{4.4}$$

$$\bar{j}_B = -D_B \nabla c_B + c_B D_B e\bar{E}/fkT \tag{4.5}$$

The condition for the electrical neutrality of the glass is

$$c_A + c_B = c_0 \tag{4.6}$$

$$\nabla(c_A + c_B) = 0 \tag{4.7}$$

where c_0 is the constant total concentration of mobile ions in the glass. The total ionic flux $\bar{j}_0$ corresponding to the electrical current density $\bar{i} = e\bar{j}_0$ is

$$\bar{j}_0 = \bar{j}_A + \bar{j}_B \tag{4.8}$$

When eqs. (4.4) and (4.5) are inserted into eq. (4.8) and eqs. (4.6) and (4.7) are used, the electric field in the glass is obtained:

$$e\bar{E}/fkT = \frac{(M - 1)\nabla c_A + M\bar{j}_0/D_A}{c_A(M - 1) + c_0} \tag{4.9}$$

The flux $\bar{j}_A$ can now be written

$$\bar{j}_A = \frac{-D_A c_0 \nabla c_A + M c_A \bar{j}_0}{c_A(M - 1) + c_0} \tag{4.10}$$

where M is the ratio of self-diffusion constants, $M = D_A/D_B$. Usually subscript A is used for the dopant ion exchanged into the glass and subscript B for the original ion in the glass. In the typical waveguide fabrication processes M is thus less than 1. The change of concentration c_A during the ion exchange, as a function of time t, can be obtained from Fick's second law:

$$\partial c_A/\partial t = -\nabla \cdot \bar{j}_A \tag{4.11}$$

With relative fluxes $\bar{J} = j_A/c_0$, $\bar{J}_0 = j_0/c_0$ and relative concentration $C = c_A/c_0$, the diffusion equation for the electric-field assisted ion exchange is

$$\partial C/\partial t = \frac{D\nabla^2 C}{C(M - 1) + 1} - \frac{D(M - 1)(\nabla C)^2 + M\bar{J}_0 \cdot \nabla C}{[C(M - 1) + 1]^2} \tag{4.12}$$

The flux $\bar{J}_0$ is proportional to the electric current density and has the dimension of speed—in fact it is the average local migration velocity of the mobile cations in the glass.

4.1.2 Equilibrium and Kinetics at the Interface of Ion Source and Glass

The binary ion exchange between salt melt and glass is described by the chemical reaction at the melt-glass interface:

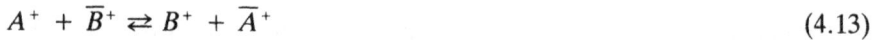

$$A^+ + \bar{B}^+ \rightleftarrows B^+ + \bar{A}^+ \tag{4.13}$$

Here A^+, B^+ and $\bar{A}^+$, $\bar{B}^+$ are the two cations in the melt phase and the glass phase, respectively. The equilibrium constant K for this reaction is

$$K = a_B \bar{a}_A / a_A \bar{a}_B \tag{4.14}$$

where a_A, a_B and $\bar{a}_A$, $\bar{a}_B$ are the thermodynamic activities of ions in the melt and in the glass, respectively. For the glass

$$\bar{a}_A / \bar{a}_B = (C_A/C_B)^\tau \tag{4.15}$$

has been used for the relation of activity ratio, with the constant τ. The salt melt can be described in terms of the regular solution theory with

$$\ln (a_A/a_B) = \ln (N_A/N_B) - (W/RT)(1 - 2N_B) \tag{4.16}$$

where N_A and N_B are the mole fractions of the two ions in the salt melt, R is the gas constant, T is the absolute temperature, and W is the interaction energy of the ions. Combining Eqs. (4.14) to (4.16),

$$\ln (N_A/N_B) - (W/RT)(1 - 2N_B) + \ln K = \tau \ln [C_A/(1 - C_A)] \tag{4.17}$$

Garfinkel [2] was able to describe the ion-exchange equilibrium by this equation. For example, in Ag^+-Na^+ exchange from a $AgNO_3/NaNO_3$ melt the process has been found not to be limited by the transfer of ions in the melt to the glass interface or by the interface reaction kinetics, because the time to reach equilibrium at the surface is very short [3]. Therefore, the concentration C_B given by eq. (4.17) can be used as a surface boundary condition. A relationship (4.17) can be found experimentally by using different melt fractions. The interaction energy W is known for usual salt mixtures. The values of τ and K depend on the glass used, for AgNO$_3$/ NaNO$_3$ melts they have been found for example in [2–4]. Generally even quite low Ag$^+$ melt fractions produce high C_{Ag} concentrations in the glass. For N_{Ag}/N_{Na} = 0.005, C_{Ag} is about 0.5 (Fig. 4.1). Similar nonlinear partitioning has been found for Cs$^+$-K$^+$ exchange [5].

When no external electric field is used, the flux $\bar{J}_B$ through the interface is balanced by the opposite flux $\bar{J}_A = -\bar{J}_B$. In field-assisted ion exchange the total flux through the interface is given by the electrical current density according to eq. (4.8), but the boundary condition remains the same.

When an Ag metal thin film is in contact with the glass, a description in terms of the exchange of ions between two media is not possible. The transfer of Ag$^+$ ions from metal into glass is by an electrochemical reaction:

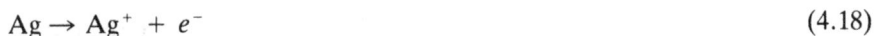

$$Ag \rightarrow Ag^+ + e^- \tag{4.18}$$

When no electric field is used, eq. (4.8) gives an equal but opposite flux of Na$^+$ ions, which must be reduced to metallic sodium as they transfer out of the glass.

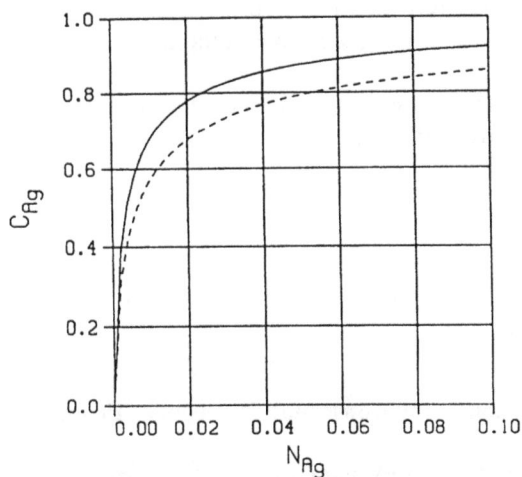

re 4.1 The dependence of relative Ag$^+$ concentration at the glass surface on the Ag$^+$ ion ratio in NaNO$_3$ melt. The curves are calculated from the results experimentally found for two soda lime glasses, in reference [4] (solid line) and [3] (dashed line).

The equilibrium of the reaction is strongly against this transfer. Experimentally, only very low Ag^+ concentrations have been produced by using Ag thin-film sources without an electric field [6]. In fact, silver ions in the glass tend to form metal colloids, especially at high concentrations.

When an electric field is used in this process, it can be described as an electrolysis of Ag^+ ions from the silver anode into the glass, which acts as a solid electrolyte. Equation (4.18) can be considered to represent the only anode reaction, so that the flux at the interface is given by

$$\bar{J}_{Ag} = \bar{J}_0 \tag{4.19}$$

Because the metal anode is a surface of constant electrical potential, this flux is always perpendicular to the surface. If this direction is chosen to be along the y-coordinate axis, the surface boundary condition is, from eq. (4.10),

$$\partial C_{Ag}/\partial y = J_0(C_{Ag} - 1)/D \tag{4.20}$$

As the silver thin film is a source of only Ag^+ ions, it is similar to the pure $AgNO_3$ melt in one respect: total exchange of Ag^+ ions for Na^+ ions can be produced. However, it takes a finite time for surface concentration to reach the maximum value, so that the difference of boundary conditions corresponds to the physical differences in the two processes: with molten salts there is both diffusion and migration in the electric field of ions from the source; with Ag film there is only migration. Experimentally it has been verified that the surface concentration of silver can be controlled with the external voltage, when Ag thin-film sources are used [7,8].

Finally, for those glass surfaces not in contact with a source for ion exchange, all ion flux components normal to the surface are 0. From eqs. (4.8) and (4.10) it then follows that

$$\partial C/\partial y = 0 \tag{4.21}$$

which is in fact a special case of condition (4.20), with no component of $\bar{J}_0$ perpendicular to the surface.

4.1.3 One-Dimensional Solutions: Slab Waveguide Processes

For slab waveguides the ion exchange is made uniformly for the entire substrate surface. This simplifies the calculation of ion concentration profiles considerably, because differential equations need to be solved in only one dimension. Slab waveguide fabrication processes provide illuminating examples on the principles of modeling ion exchange and the general properties of ion concentration profiles.

There are two basic categories of ion exchange processes. In thermal diffusion processes the dominating transport mechanism for ions in the glass is diffusion. In field-assisted processes the transport by migration in the electric field caused by external voltage prevails.

In one-dimensional field-assisted ion-exchange processes the electric current density J_0 is position independent as there are no current sources or sinks inside the glass. The electric field in the glass is then given by the eq. (4.9). The electric potential difference between the glass surfaces, with substrate thickness d and the origin of x-axis at the glass surface,

$$U = \int_0^d E \, dx \tag{4.22}$$

must have the correct value U used as the external voltage in the process. The current density can be solved from eqs. (4.9) and (4.22). Usually the exchanged ions have different mobilities, and the total resistance of the glass will change during the process. However, this change is often small, because only the shallow surface layer is affected, and electric current density that is constant in time can be used in calculations.

A one-dimensional form of the eq. (4.12) is

$$\partial C/\partial t = \frac{\partial}{\partial x} \left[\frac{D \, \partial C/\partial x}{C(M - 1) + 1} \right] - \frac{M J_0 \partial C/\partial x}{[C(M - 1) + 1]^2} \tag{4.23}$$

The first term on the right includes the effect of diffusion; the second term, the effect of migration due to the electric current flow through the substrate. Especially, for thermal ion exchange where no external current source is used, the equation is obtained in the form

$$\partial C/\partial t = \frac{\partial}{\partial x} \left[\frac{D \, \partial C/\partial x}{C(M - 1) + 1} \right] \tag{4.24}$$

If the two ions have same mobilities, $M = 1$ and eq. (4.23) is simplified:

$$\partial C/\partial t = D \, \partial^2 C/\partial x^2 - J_0 \partial C/\partial x \tag{4.25}$$

For thermal ion exchange a simple diffusion equation is obtained with concentration-independent rates of diffusion and migration:

$$\partial C/\partial t = D\, \partial^2 C/\partial x^2 \tag{4.26}$$

Equations (4.25) and (4.26) are also obtained at the limit of low concentration, $C \ll 1$. This is the case for typical ion exchange with a diluted molten salt source to limit the exchanged concentration to a low value.

For a simple thermal diffusion, the initial and boundary conditions are

$$\begin{aligned} C(x, 0) &= 0 \\ C(0, t) &= C_0 \end{aligned} \tag{4.27}$$

and eq. (4.26) has the analytic solution

$$C(x, t) = C_0\, \mathrm{erfc}\,[x/(2\sqrt{Dt})] \tag{4.28}$$

with the complementary error function

$$\mathrm{erfc}(z) = \frac{2}{\sqrt{\pi}} \int_{z}^{\infty} e^{-t^2}\, \mathrm{d}t \tag{4.29}$$

Often it is difficult to find out separately the values of both D and M from experiments. Then it can be convenient to approximate the diffusion with eq. (4.25) using an effective diffusion constant value for D.

The full field-assisted eq. (4.23) is used to examine slab waveguide fabrication by ion migration from a surface source. In a typical process the boundary condition at the surface is $C(0, t) = 1$, and $M < 1$, so that the dopant ions migrating into the glass have smaller mobility than the original ions. Through balance of diffusion and migration, this case leads to the formation of the stationary-shape profile penetrating deeper into the substrate with the speed J_0 [9]. For the profile with a constant shape in the reference frame moving with speed J_0,

$$\partial C/\partial t = -J_0 \partial C/\partial x \tag{4.30}$$

which gives the stationary-shape condition: the flux of the ions is everywhere proportional to their concentration

$$J = \frac{-D\partial C/\partial x + MJ_0 C}{C(M - 1) + 1} = J_0 C \tag{4.31}$$

where eq. (4.10) for the flux was used. The solution to this differential equation is

$$C = \{1 + \exp[J_0(1 - M)(x - J_0t)/D]\}^{-1} \tag{4.32}$$

It takes some time for the concentration distribution to reach the stationary migration state. One condition is that eq. (4.32) must give the correct boundary condition at the surface. This gives the approximate limit for the time necessary to reach this state:

$$t > 5D/[J_0^2(1 - M)] \tag{4.33}$$

With equal mobilities of the two ions, $M = 1$, a stationary profile will not develop. In this case, the solution is from [10]:

$$C = 0.5\left[\text{erfc}\left(\frac{x - J_0t}{2\sqrt{Dt}}\right) + \exp\left(\frac{J_0x}{D}\right)\text{erfc}\left(\frac{x + J_0t}{2\sqrt{Dt}}\right)\right] \tag{4.34}$$

For deep penetration, the first term dominates and the second one can usually be ignored. The profile is again of the complementary error function-type, shifted to the depth J_0t.

For more general solutions numerical solving methods for eq. (4.23) must be used. These are needed to calculate anything but the stationary profile when the mobilities of the two ions are unequal. Also, numerical methods allow the modeling of processes with nonconstant boundary conditions, including all two-step processes.

For solution of diffusion equations finite difference methods are most powerful. As an example the partial derivatives in the eq. (4.26) can be exchanged for finite difference expressions:

$$\frac{C_i^{n+1} - C_i^n}{\Delta t} = D\frac{C_{i+1}^n - 2C_i^n + C_{i-1}^n}{h^2} \tag{4.35}$$

Here, concentration is written in a regular array of points x_i with spacing h between neighboring points. Calculation is also performed for discrete values of time t^n with interval Δt between successive calculated concentrations. C_i^n is concentration at point x_i in time t^n. From eq. (4.35) concentration at the next time step can be explicitly solved:

$$C_i^{n+1} = C_i^n + \frac{D\Delta t}{h^2}(C_{i+1}^n - 2C_i^n + C_{i-1}^n) \tag{4.36}$$

This simple method of solution is called the *Euler method*. The disadvantage of the Euler method is its instability, which necessitates the use of small time intervals, Δt.

A somewhat more elaborate finite difference method is the Dufort-Frankel method [11]. This uses eq. (4.26) written in form

$$\frac{C_i^{n+1} - C_i^{n-1}}{2\Delta t} = D \frac{C_{i+1}^n - (C_i^{n+1} + C_i^{n-1}) + C_{i-1}^n}{h^2} \tag{4.37}$$

Again it is simple to solve C_i^{n+1} in explicit form. The Dufort-Frankel method uses two successive levels, $n - 1$ and n, to calculate the next level, $n + 1$. To start the calculation one level has to be calculated with some other method; for example, with the Euler method. Dufort-Frankel method allows the use of longer time steps, Δt, so that calculation is faster than with the Euler method.

For eq. (4.23), the Dufort-Frankel type finite-difference form is

$$\frac{C_i^{n+1} - C_i^{n-1}}{2\Delta t} = D \frac{C_{i+1}^n - (C_i^{n+1} + C_i^{n-1}) + C_{i-1}^n}{[C_i^n(M - 1) + 1]h^2}$$

$$- D \frac{(C_{i+1}^n - C_{i-1}^n)^2(M - 1)}{[C_i^n(M - 1) + 1]^2 4h^2} - \frac{MJ_0(C_{i+1}^n - C_{i-1}^n)}{[C_i^n(M - 1) + 1]^2 2h} \tag{4.38}$$

Figure 4.2 compares two thermal ion-exchange profiles, calculated with values $M = 1.0$ and $M = 0.1$ and all other parameters the same. The initial and boundary conditions are those given by (4.27). The distribution calculated with value $M = 1.0$ is the same as given by the eq. (4.28). With value $M = 0.1$ the rate of diffusion is increased at higher concentrations—this increases the amount of exchange and decreases the concentration gradient close to surface.

Figure 4.3(a) compares the analytical and the numerically calculated distributions for field-assisted ion exchange with value $M = 1.0$. Figure 4.3(b) shows a similar comparison for the value $M = 0.1$. Again, the initial and boundary conditions are given by (4.27). We see that the numerical results give concentration profiles that have penetrated deeper than the analytically obtained depth. In fact this is expected, because with depth $J_0 t$ both eqs. (4.32) and (4.34) give the total relative amount of dopant ions per area equal to $J_0 t$, which is the quantity carried by the electrical current into the glass. As discussed in Section 4.1.2, there is also diffusion from the source into the glass until zero concentration gradient is reached at the surface. This gives the extra amount of dopant ions in the glass. In fact, with a silver thin-film source the no-diffusion boundary condition, eq. (4.20), is used, and the analytic solutions are reproduced by numerical calculation.

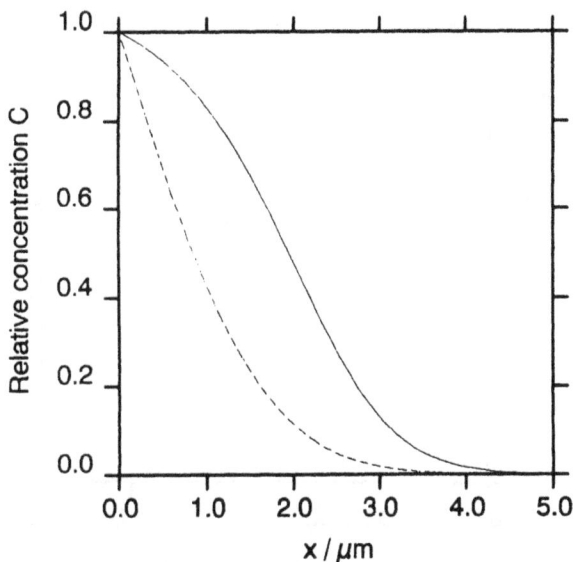

Figure 4.2 Profiles of relative concentration C calculated for thermal ion exchange of slab waveguides: $M = 0.1$ (solid line) and $M = 1.0$ (dashed line); other parameters are $D = .002 \ \mu m^2/s$, $t = 100$ s, and $J_0 = 0 \mu m/s$.

4.1.4 Two-Dimensional Solutions: Channel Waveguide Processes

For channel waveguides, ion exchange is limited to a stripe region on the glass surface. Two-dimensional concentration distributions are obtained. There is a general mirror symmetry for these waveguides, which will make it necessary to calculate the distributions only on one-half of the waveguide cross section.

To understand field-assisted fabrication of channel waveguides, the two-dimensional electric current distribution in the glass during these processes must first be examined. In principle, the electric field and current distributions can be calculated from the conductivity distribution of the glass with the boundary conditions of constant electric potentials at the electrode surfaces. However, recalculation of the field again and again, as the glass conductivity changes during the ion exchange, would be very time consuming. Obviously, some simplification is necessary. As in the one-dimensional situation, we can expect the change in current distribution to be small, because ion exchange occurs only in a shallow region under the source stripe. The electric current distributions of homogeneous glass with the given electrode geometry may be used. The electric-field distributions in analytic form for stripe anode geometries are given in [12]. For a single-stripe geometry

(a)

(b)

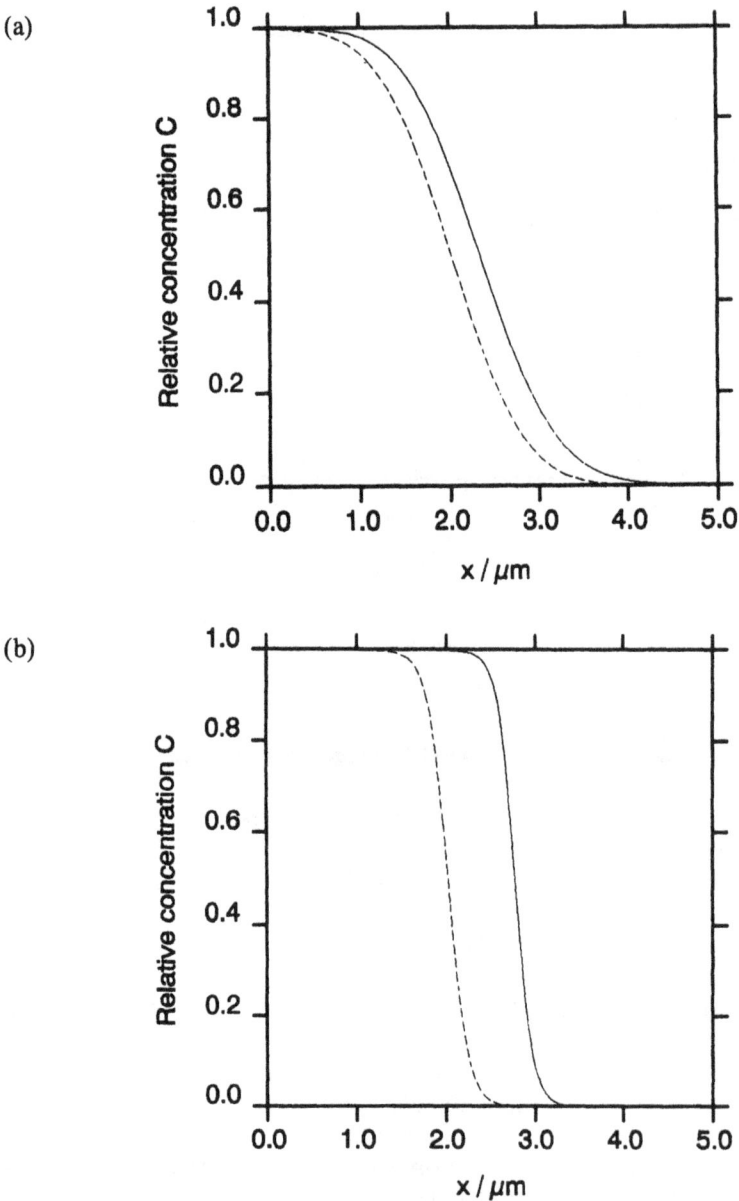

Figure 4.3 Profiles of relative concentration C calculated for field-assisted ion exchange of slab wave-guides. Numerical solutions are drawn with solid curves and analytical solutions with dashed curves: $M = 1.0$ (a) and $M = 0.1$ (b); other parameters are $D = .002$ $\mu m^2/s$, $t = 100$ s, and $J_0 = 0.02$ $\mu m/s$.

this distribution is illustrated in Figure 4.4. The electric-field distribution is for a substrate glass of thickness d, stripe width w, and electrode voltage U:

$$\overline{E} = E_x + iE_y = i(\pi/2) \frac{U}{dK(m)} \left[\frac{\tanh^2(u) - 1}{\tanh^2(u) - m^2} \right]^{1/2} \tag{4.39}$$

Here $i = \sqrt{-1}$, $m = \tanh(\pi w/4d)$, $u = (\pi/2)(x - jy)/d$, and $K(m)$ is

$$K(m) = \int_0^{\pi/2} \frac{d\theta}{[m^2 + (1 - m^2) \sin^2 \theta]^{1/2}} \tag{4.40}$$

For a regular array of stripes with the axial distance of g between the stripes $(g \ll d)$, the electric-field distribution is

$$\overline{E} = E_x + jE_y = j \frac{U}{dK(m)} \left(\frac{\sin^2(u) - 1}{\sin^2(u) - m^2} \right)^{1/2} \tag{4.41}$$

with $m = \sin(\pi w/2g)$, $u = \pi(x - jy)/g$, and

$$K(m) = \int_0^{\pi} \left(\frac{\sinh^2(\Theta d/b) + 1}{\sinh^2(\Theta d/b) + m^2} \right)^{1/2} d\Theta \tag{4.42}$$

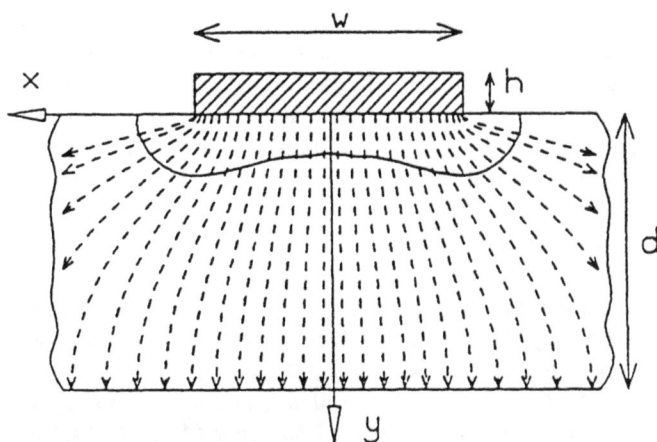

Figure 4.4 Electric-field distribution in glass with a single-strip anode; also shown is the waveguide boundary migrating into the substrate.

The field distributions given by eqs. (4.39) and (4.41) are solutions of Laplace's equation and therefore have zero divergence. The flux distribution is obtained from the field just by multiplication:

$$\bar{J}_0 = \sigma \bar{E}/ec_0 \tag{4.43}$$

where σ is given by the Nernst-Einstein relation (4.3). The flux distribution fulfills the continuity condition $\nabla \cdot \bar{J}_0 = 0$ and the boundary conditions of constant electrode potentials. Again, the correct potential difference across the substrate can be obtained by using eq. (4.22), with integration along the $x = 0$ axis. This will change the flux distribution $\bar{J}_0$ only by a linear relation.

No accurate analytic solutions exist for channel waveguide concentration distributions. However, approximate profiles for field-assisted waveguides can be obtained without the full solution of eq. (4.12). Here analogy with one-dimensional solutions is used. Both eqs. (4.32) and (4.34) represent nearly step-index slab waveguides with the waveguide depth $J_0 t$. In other words, during the ion migration, the boundary of the waveguide is penetrating into the substrate with speed J_0. In the field-assisted fabrication process for a channel waveguide, the migrating ions start to penetrate into the substrate glass from the source stripe and the waveguide boundary will move along the lines of electrical flux with local velocity $\bar{J}_0(x, y)$. The waveguide boundary curve is given by the points

$$\bar{r}_b(x_0, t) = \bar{r}_0 + \int_0^t \bar{J}_0(\bar{r}) \, dt \tag{4.44}$$

with $\bar{r}_0 = (x_0, 0)$, the starting point coordinates at the stripe. This approximation is quite good when penetration depth is substantially larger than the diffusion length. The graded-index nature of the boundary can be obtained for deep waveguides, too, as the concentration profiles may be assumed to have achieved locally stationary shapes, with concentration gradient parallel to the flux and given near the boundary point $\bar{r}_b$ according to eq. (4.31):

$$C(\bar{r}) = \{1 + \exp[\bar{J}_0(\bar{r}_b) \cdot (\bar{r} - \bar{r}_b)(1 - M)/D]\}^{-1} \tag{4.45}$$

Figure 4.5 shows the calculated concentration distribution in Corning 0211 glass from a 250 s long field-assisted silver ion exchange with single stripe width $w = 30$ μm, voltage $U = 10$ V. Material parameters are $D = .0020$ $\mu m^2/s$, $M = 0.7$, $f = 0.51$, $c_0 = 5.0 \times 10^{-3}$ mol/cm^3, and $d = 0.6$ mm. This waveguide profile was calculated from eqs. (4.39), (4.40), and (4.43) to (4.45).

To model more general channel waveguide processes, finite difference methods in two dimensions must be used [13,14]. The two-dimensional eq. (4.12) is

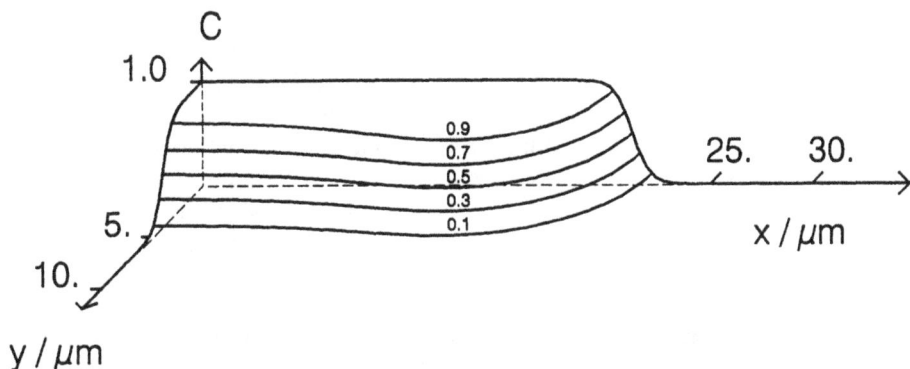

Figure 4.5 A three-dimensional plot of the calculated relative concentration distribution $C(x, y)$ of a channel waveguide fabricated by a 250 s long field-assisted silver ion exchange with stripe width $w = 30$ μm and voltage $U = 10$ V; the material parameters are $D = .0020$ μm^2/s, $M = 0.7, f = 0.51, c_0 = 5.0 \times 10^{-3}$ mol/cm^3, and $d = 0.6$ mm.

changed to a finite-difference form. Material parameters, proper boundary conditions, and the electric current distribution in the glass define the process. Because a great deal of calculation is usually involved, the choice of the calculation area and the two-dimensional grid points is important to optimize the time needed for computing.

4.2 THE REFRACTIVE INDEX CHANGE FROM ION EXCHANGE

Refractive indices of glass compositions with oxide constituents can be calculated from a model given in [15] and [16]. This model, based on Gladstone-Dale relations, uses a number of constants to calculate from the weight fractions of the different oxide constituents of the glass first the density of the glass and then the refractive index as a function of wavelength. The model actually gives a range for the density and refractive index, because the value of these quantities depends not only on the composition but also on the annealing history of the glass.

Applied to ion-exchanged waveguides, the model predicts an approximate linear relation between the relative exchanged concentration C_A and the refractive index change Δn_0:

$$\Delta n_0 = \frac{C_{Ag}}{V_0}\left[\Delta R - \frac{R_0 \Delta V}{V_0}\right] \tag{4.46}$$

Here V_0 and R_0 are, respectively, the volume of glass per gram of oxygen atoms and the refraction per gram of oxygen atoms in the original composition. The terms ΔV and ΔR are the changes of these quantities resulting from the total replacement of the original ions by the dopant ions. Therefore, there are two separate contributions to the refractive-index change: from the change of ionic polarizability and from the volume change of the glass. Typically, in waveguide fabrications, refractive-index increasing ions with higher ionic polarizability also have higher ionic radii, which would make the glass expand. The first contribution to the refractive-index change would be positive; the second, negative. Because the ion exchange affects only the surface of the glass, the expansion is limited. As glass can expand only in the direction normal to the surface, it stays compressed from its free volume.

An alternative way of looking at the deviation from formula (4.46) caused by the compression is as a stress-optical effect. For a slab waveguide, compressive stress s exists in the directions parallel to the surface, whereas in the direction normal to the surface there is no stress because of the free expansion. This nonisotropic stress causes birefringence, and the index changes for the two polarizations are [17]

$$\Delta n_{TE} = \Delta n_0 + (C_1 + C_2)s$$
$$\Delta n_{TM} = \Delta n_0 + 2C_2 s$$

$$(4.47)$$

where C_1 and C_2 are the elasto-optic coefficients. Often only the birefringence factor, Brewster's constant $B = C_2 - C_1$, is known. Albert and Yip [18] have shown that for K^+-Na^+ exchange the refractive index increase is due mainly to induced stress. They also calculated the value for birefringence in potassium-exchanged waveguides, which was close to the observed value. For silver-exchanged waveguides they predicted birefringence on the order of 5×10^{-4}, which is approximately the value experimentally observed. For channel waveguides, the stress distribution is more complex.

It is usually not possible for a given glass to accurately reflect the purely theoretical relationship between the concentration distribution and the refractive index increase. However, with good accuracy this relationship is linear, with the proportionality factor that depends on the polarization and optical wavelength. With the proportionality factor determined experimentally for a given wavelength, it can be extrapolated to other wavelengths for the known glass composition.

4.3 MODELING OPTICAL PROPAGATION IN ION-EXCHANGED WAVEGUIDES

The purpose of this section is to introduce basic tools for modeling the optical propagation in waveguides with known refractive-index profiles. The described methods are suited for graded-index glass waveguides. First, just a brief discussion of the theory of waveguide modes is given. This is not meant as an introduction to this subject, but to review the general properties of the waveguide modes. For an introduction, any textbook on optical waveguide theory, for example [19], can be used.

Again, the slab waveguides are studied first to form a basis for modeling more complicated channel waveguides. The modeling of slab waveguides is based on the powerful WKB-approximation, which can be used for calculating the mode propagation constants for the known waveguide profile. Alternatively, inverse WKB methods give approximate waveguide profiles from the measured propagation constants. With the waveguide profile and the mode propagation constants, the transverse mode field distributions also may be solved.

For modeling the channel waveguides, the widely used effective index method is first described. This approximative method applies the slab waveguide solving methods directly to channel waveguide problems.

The next subsection is devoted to description of finite difference and finite element methods, which are used for solving the modes of channel waveguides. In particular, an efficient finite difference method is described in detail. Finally, a powerful tool for modeling optical propagation in three-dimensional waveguide structures, the beam propagation method, is introduced.

4.3.1 Properties of Optical Waveguide Modes

An optical waveguide is characterized by the refractive-index distribution $n(x, y)$, which is here taken to be scalar, real, and independent of the z-coordinate, because the z-axis is chosen to be along the propagation direction in the waveguide. For slab waveguides, $n(x)$ is a function of x-coordinate only. The waveguide is surrounded by the substrate and cladding materials, and the refractive index in the waveguide is higher than in the surroundings. The translational symmetry of the waveguide along the z-axis makes possible the examination of optical propagation in terms of the optical modes of the structure. These are defined as electromagnetic-field solutions of the type

$$\overline{E}(x, y, z) = \overline{E}_\Gamma(x, y) \exp(-j\beta_\Gamma z)$$
$$\overline{H}(x, y, z) = \overline{H}_\Gamma(x, y) \exp(-j\beta_\Gamma z)$$

(4.48)

for electric and magnetic field $\overline{E}$ and $\overline{H}$, with mode label Γ, and the propagation constant of the mode β_Γ. Due to the additional symmetry in slab waveguides, mode fields, too, have no y dependence.

The guided modes have their field energy localized in the vicinity of the waveguide. For a given wavelength, λ, there is a limited number of guided modes with discrete values of propagation constant $\beta = k_0 n_{\text{eff}}$, where $k_0 = 2\pi/\lambda$ and n_{eff} is the effective index of the mode, which is higher than the refractive index of either the surrounding substrate or the cladding. The harmonic time dependence of the field—$e^{\omega t}$, where $\omega = k_0 c$, with the vacuum speed of light c—has been left out of (4.48). In addition to the guided modes, there is a continuum of radiation modes, which are not bound to the waveguide.

The orthogonality relation of the modes is

$$\iint_{-\infty}^{+\infty} \mathrm{d}x\mathrm{d}y\ \overline{E}_{t\Gamma} x \overline{H}_{t\mu}^* = 0, \qquad \beta_{\Gamma} \neq \beta_{\mu} \tag{4.49}$$

Here label t indicates the field component is transverse to the z-direction. In addition to the orthogonality, the modes are also a complete set, so that an arbitrary field distribution can be expressed as a superposition of guided and radiation modes, labeled by Γ and μ, respectively:

$$\overline{E}_t(x, y) = \Sigma\ a_{\Gamma}\overline{E}_{\Gamma}(x, y) + \int \mathrm{d}\mu\ a(\mu)\overline{E}(\mu; x, y)$$
$$\overline{H}_t(x, y) = \Sigma\ a_{\Gamma}\overline{H}_{\Gamma}(x, y) + \int \mathrm{d}\mu\ a(\mu)\overline{H}(\mu; x, y) \tag{4.50}$$

Expansion coefficients a_{Γ} can be calculated with the orthogonality relation (4.49).

An example of the use of mode orthogonality properties is the calculation of optical coupling loss in a butt coupling from a single-mode fiber into a single-mode channel waveguide. This loss includes both reflection and mode mismatch losses. The mode mismatch loss consists of the portion of light transmitted through the fiber-waveguide interface that is coupled into radiation modes. The fraction of transmitted light coupled into the waveguide mode is given by the overlap integral

$$I = \frac{\left(\iint_{-\infty}^{\infty} \mathrm{d}x\mathrm{d}y\ \overline{E}_{f1} x \overline{H}_{f2}^*\right)^2}{\left(\iint_{-\infty}^{\infty} \mathrm{d}x\mathrm{d}y\ \overline{E}_{f1} x \overline{H}_{f1}^*\right)\left(\iint_{-\infty}^{\infty} \mathrm{d}x\mathrm{d}y\ \overline{E}_{f2} x \overline{H}_{f2}^*\right)} \tag{4.51}$$

Here $\overline{E}_{f1}, \overline{H}_{f1}$ and $\overline{E}_{f2}, \overline{H}_{f2}$ are the transverse electric and magnetic field distributions of the fiber mode and the waveguide mode, respectively.

4.3.2 WKB Analysis of Slab Waveguides

4.3.2.1 Calculation of Mode Propagation Constants

In slab waveguides two mode sets with different polarization properties exist. The TE-modes have electric field vectors in the y-direction, transverse to the directions of propagation and the surface normal. The TM-modes have magnetic field vectors in this direction.

For calculation of propagation constants in graded-index slab waveguides, WKB approximation (Wentzel-Kramers-Brillouin approximation, a method used in quantum mechanics to solve the Schrödinger equation) is generally used. Here the basis of the method is described with a ray-optic picture of light propagation in waveguide.

The waveguide profile shown in Figure 4.6 has a step-index boundary at $x = 0$ and a graded-index distribution that decreases with depth. This is a typical profile obtained by ion exchange through the substrate surface. Also shown is path of a ray propagating in the waveguide by total internal reflections. Because the propagation constant $\beta = k_0 n_{eff}$ is the z-component of the ray wave vector, the transverse x component is

$$k_x = k_0[n^2(x) - n_{eff}^2]^{1/2} \tag{4.52}$$

Therefore the ray path can be obtained directly when the propagation constant is known. At the depth x_a, where

$$n(x_a) = n_{eff} \tag{4.53}$$

the transverse wave vector component is 0. This is the turning point of the ray.

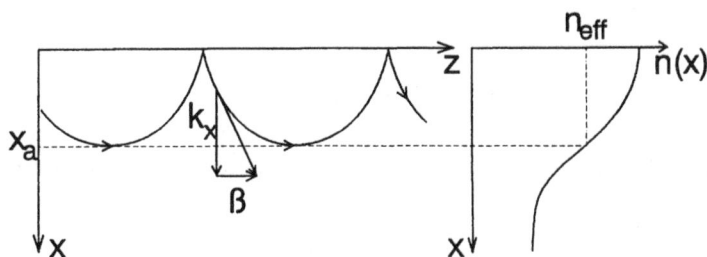

Figure 4.6 The ray-optic picture of light propagation in a slab waveguide.

For the usual application of the WKB method, the ray path must be limited to a graded-index region of the slab waveguide. The limits of the ray path are the two turning points, outside which the refractive index is lower than the effective index of the ray.

For the ray to correspond to a guided mode, the total phase shift for a round trip must be a multiple of 2π:

$$2k_0 \int_0^{x_a} [n^2(x) - n_{\text{eff}}^2]^{1/2} \, dx - \Phi_1 - \Phi_2 = 2m\pi \tag{4.54}$$

Here $m = 0, 1, 2, \ldots$, and Φ_1 and Φ_2 are the phase shifts from total internal reflection at the turning points. For the step-index interface, the phase shift is

$$\Phi = 2 \arctan\{p[(\beta^2 - k_0^2 n_c^2)/(k_0^2 n_f^2 - \beta^2)]^{1/2}\} \tag{4.55}$$

where $p = 1$ for TE-polarization and $p = (n_f/n_c)^2$ for TM-polarization. The terms n_f and n_c are the refractive indices at the inside and outside of the step, respectively.

With the turning-point at the graded-index profile, the phase shift is obtained as a limit value:

$$\Phi = \pi/2 \tag{4.56}$$

Equation (4.54) with phase shifts (4.55) and (4.56) is the WKB relation. The propagation constants of modes with mode indices m can be obtained, when the refractive index profile $n(x)$ is known. This usually has to be done with numerical integration.

As an example, Figure 4.7 shows the results for calculated propagation constants for diffused waveguides with refractive index profiles from eq. (4.28):

$$n(x) = n_c, \qquad x < 0$$
$$n(x) = n_s + (n_f - n_s) \, \text{erfc}[x/(2\sqrt{Dt})], \qquad x > 0 \tag{4.57}$$

For the sake of generality, normalized parameters are used. The normalized frequency V is defined

$$V = k_0 2\sqrt{Dt}(n_f^2 - n_s^2)^{1/2} \tag{4.58}$$

The normalized propagation constant is

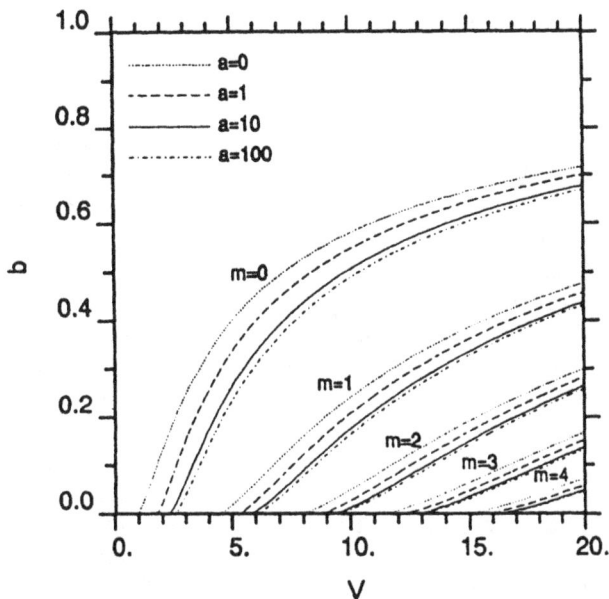

Figure 4.7 Normalized propagation constants b for TE-modes in diffused slab waveguides as a function of normalized frequency V, with different asymmetries a.

$$b = (n_{\text{eff}}^2 - n_s^2)/(n_f^2 - n_s^2) \qquad (4.59)$$

and the asymmetry factor is

$$a = (n_s^2 - n_c^2)/(n_f^2 - n_s^2) \qquad (4.60)$$

The mode propagation properties for diffused slab waveguides, which can be characterized with a single diffusion constant, may be obtained from the figure.

4.3.2.2 Inverse WKB Analysis

Often in practical situations experimental propagation constants, typically measured with prism coupling, are known. Therefore the need arises to invert the WKB relation (4.54) to get the refractive index profile of the slab waveguide. First, it must be emphasized that though there exists only one set of propagation constants for a known waveguide at a given wavelength, there are different waveguide profiles with the same set of discrete propagation constants. Therefore, when inverse WKB analysis is made, some functional form for the index profile must be assumed. One possibility is to represent the refractive index distribution with a simple function of a few unknown parameters and find the values for parameters that best fit the propagation constants.

If the only known factors of the waveguide profile shape are its graded index and that it decreases with depth from the surface, we may assume it to be linear between turning points of the different modes. The mode effective indices now give the refractive-index values at the turning points, so the only unknowns are the depths of the turning points and the maximum refractive index at the surface. Assuming a certain value for the maximum refractive index at the surface and proceeding from the lowest-order mode to the highest-order mode, the turning point depths can be solved from (4.54) for one mode at a time. If the surface refractive index is unknown, a value for surface index may be found that minimizes the total curvature of the waveguide profile.

A more sophisticated inverse WKB method has been introduced by Chiang [20]. The measured effective indices are plotted as a function of mode index m, and a curve is fitted to these data. The surface index is given by extrapolation to the value $m = -0.75$. To obtain smooth waveguide profiles, virtual modes with fractional mode indices are introduced. The effective indices of these virtual modes are read from the fitting curve. With this artificial increase in the number of the turning points a very smooth graded-index distribution is attained.

4.3.2.3 Mode Field Distribution of Slab Waveguides

The mode field distribution in a graded-index slab waveguide is a solution of the wave equation

$$\mathrm{d}^2\psi/\mathrm{d}x^2 = (\beta^2 - n(x)^2 k_0^2)\psi \tag{4.61}$$

Here ψ is the field component E_y for TE-modes and H_y for TM-modes. In addition there are boundary conditions at the interfaces that demand the continuity of E_y and $\mathrm{d}E_y/\mathrm{d}x$ for the TE-modes and the continuity of H_y and $n^{-2}\mathrm{d}H_y/\mathrm{d}x$ for the TM-modes.

Several methods exist for finding the solutions to (4.61). These solutions are the propagation constants and the field distributions of the modes. The WKB method gives solutions for the propagation constants β, so it is convenient to use these values to find out the field distributions by solving the differential equation (4.61).

This kind of method for surface waveguides has been described by Ramaswamy and Lagu [21]. To start the calculation for the solution of differential equation (4.61), the value of field and the slope of field at a single point is needed.

Because for guided modes the decaying solution for the field in the uniform cladding is

$$\psi(x) = \psi(0)e^{\alpha x}, \qquad x < 0 \tag{4.62}$$

where

$$\alpha = (\beta^2 - k_0^2 n_c^2)^{1/2} \tag{4.63}$$

it is given by the boundary conditions at the surface for a TE-mode:

$$d\psi/dx = \alpha\psi(0), \qquad x = 0 \tag{4.64}$$

As the mode field solution can be multiplied by an arbitrary constant, $\psi(0)$ may be taken to have unit value, $\psi(0) = 1$. The numerical solution to (4.61) for $x > 0$ can now be found starting from $x = 0$ by a Runge-Kutta method. The Runge-Kutta method gives the field value at a point $x_{n+1} = x_n + h$ (with small step h) from the known values $\psi(x_n) = E_n$, $d\psi/dx = E_n'$ at x_n:

$$E_{n+1} = E_n + h[E_n' + (k_1 + 2k_2)/6]$$
$$E_{n+1}' = E_n' + k_1/6 + 2k_2/3 + k_3/6 \tag{4.65}$$

with

$$k_1 = h[\beta^2 - k_0^2 n^2(x_n)]E_n$$
$$k_2 = h[\beta^2 - k_0^2 n^2(x_n + h/2)](E_n + hE_n'/2 + hk_1/8)$$
$$k_3 = h[\beta^2 - k_0^2 n^2(x_n + h)](E_n + hE_n' + hk_2/2)$$

However, solutions calculated with this method usually do not correspond to the guided modes, because they do not decay to 0 at long distances from the waveguide. This numerical instability is a consequence of a slight inaccuracy in calculated propagation constant values. To obtain the guided mode field, another solution is calculated with a small change in the starting slope $d\psi/dx$ at $x = 0$. Then a linear combination of these two solutions is made that reaches 0 at a depth chosen far from the waveguide. This superposition of two solutions is the field distribution of the mode.

As an example, TE-mode fields in the profile given by eq. (4.57), with $n_s = 1.52$, $n_f = 1.527$ and $\sqrt{Dt} = 10 \ \mu m$, for 1.3 μm wavelength, are shown in Figure 4.8.

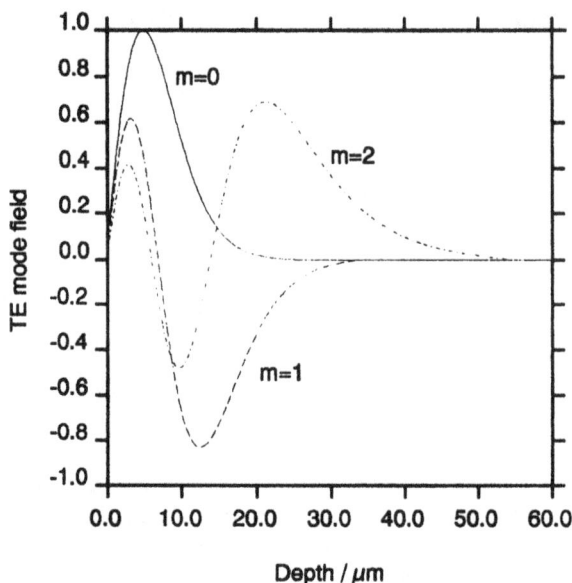

Figure 4.8 Electric field distributions of TE-modes in a complementary error function-type refractive-index profile: $n(x) = 1.520 + 0.007$ erfc $(x/20 \ \mu m)$, with wavelength 1.3 μm.

4.3.3 Effective-Index Method

The effective-index method applies the slab-waveguide solving tools to channel waveguide problems. The approximate propagation constants in a waveguide with two-dimensional refractive-index profile $n(x, y)$ are calculated by solving one-dimensional waveguide problems separately in the two dimensions.

Usually the channel waveguides are situated at the substrate surface or near it. This gives the natural choice of the coordinate axes and also induces splitting into quasi-TE and quasi-TM polarization modes. The x coordinate is chosen to give depth from the surface, just as in slab waveguides. The modes are now labeled with two indices: as TE_{ij} and TM_{ij}, with i and j as the number of nodes in transverse field distribution in x and y directions, respectively.

The principle of the effective-index method is illustrated in Figure 4.9. When the y coordinate is fixed to the value y_0, the index distribution as a function of depth x, $n(x, y_0)$, is a slab waveguide profile, for which the propagation constants may be solved. These give the discrete values of effective indices $N_i(y_0)$, with mode indices i. With such solutions for all values of y_0, one-dimensional effective-index distributions $N_i(y)$ are obtained. The effective-index distributions can be used as slab waveguide profiles in y direction. When the effective-index distribution $N_i(y)$

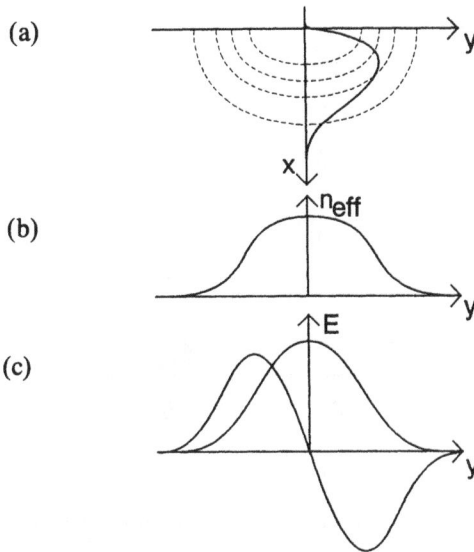

Figure 4.9 The principle of the effective index method. Propagation constants are calculated in slab profiles that are vertical slices of the channel waveguide refractive index distribution (a). This yields a horizontal effective index distribution (b). Slab waveguide modes of the effective index profile are then solved (c).

is calculated as TE-mode solutions, the propagation constants of the TM solution in this effective-index profile labeled by mode index j are the approximate propagation constants of the TE_{ij} mode in the original channel waveguide.

As an example, Figure 4.10 shows the region of single-mode operation for general diffused channel waveguides. The normalized parameter V is the normalized frequency of a slab waveguide made by same diffusion parameters, given by eq. (4.58). The other parameters are the width w of the diffusion source stripe and the diffusion depth $d = 2\sqrt{Dt}$.

The effective-index method is relatively uncomplicated and a very fast method for modeling the channel waveguides. However, more accurate methods are often needed.

4.3.4 Methods for Solving Mode Properties in Channel Waveguides

Many different methods may be used to solve the mode propagation constants and field distributions in channel waveguides. More complex methods give the full vector field solutions. However, scalar methods, which make the approximation

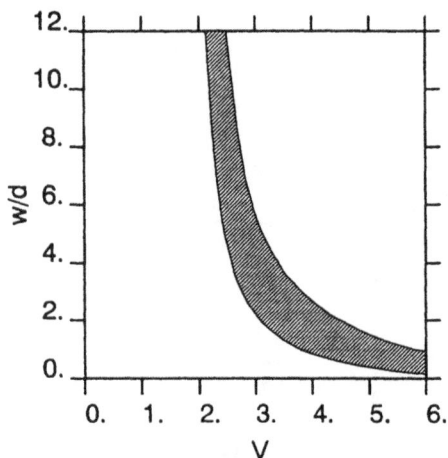

Figure 4.10 The region for single-mode operation of diffused channel waveguides is shown as a shaded area, with normalized frequency V, mask opening width w, and diffusion depth d.

of taking into account only one field component, are less complicated, faster, and in graded-index glass waveguides with small refractive-index differences, usually accurate enough. This brief description focuses on scalar methods in order to make the principles involved easier to understand.

The scalar wave equation is

$$\partial^2 E/\partial x^2 + \partial^2 E/\partial y^2 = \{\beta^2 - [k_0 n(x, y)]^2\}E \tag{4.66}$$

the mode solutions to which are of the form

$$E(x, y, z) = E(x, y) \, e^{i(\omega t - \beta z)} \tag{4.67}$$

The variational expression for propagation constant, the Rayleigh quotient, for real-valued E, is

$$\beta^2 = \frac{\displaystyle\iint \{\partial^2 E/\partial x^2 + \partial^2 E/\partial y^2 + [k_0 n(x, y)]^2 E\}E \; dxdy}{\displaystyle\iint E^2 \; dxdy} \tag{4.68}$$

The guided mode field distributions give stationary extreme values for (4.68). In particular, the fundamental mode field distribution maximizes the Rayleigh quotient.

When the concentration distribution of an ion-exchanged waveguide has been modeled by a finite-difference method and then transformed to the refractive-index profile, this profile is given by discrete values of refractive index in points of a rectangular grid. The wave equation or variational forms derived from it can be written also in finite-difference form for these grid points. This is the basis for the use of a finite-difference method for solving the mode properties [22]. If the conditions for the Rayleigh quotient to be stationary are written by differentiating the finite-difference variational expression with respect to field values at all the grid points and the result set to 0, a matrix eigenvalue equation results. Solving this matrix equation gives the mode solutions: propagation constants from the eigenvalues and field distributions from the eigenvectors.

Closely related are the scalar finite element methods [23]. The cross-sectional region of the waveguide, on which the mode field solution is to be found, is divided into smaller subregions. These elements, which fill the calculation area, are usually taken to be triangular in shape; and they can be different from each other in size. The refractive index is given a constant value within each element. The field inside each element is written as a simple interpolation function between the field values at the element corner points. Thus, the integrals in the variational expression for the field can be calculated and written in terms of the values at the discrete corner points. This expression is then differentiated with respect to all the discrete-field value parameters and the result set to 0. This again gives a matrix eigenvalue equation with the waveguide modes as solutions. The finite element methods are more efficient but also more complicated than the finite difference methods. Figure 4.11

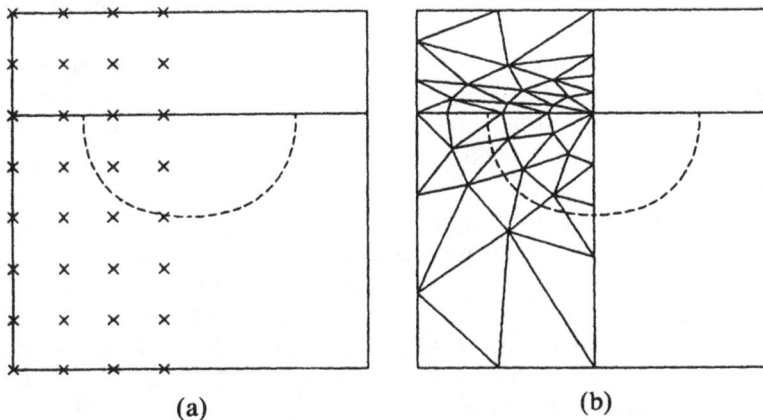

(a)　　　　　　　　　(b)

Figure 4.11 Calculation area for mode solving is chosen to include one-half a channel waveguide cross section: for the finite-difference method (a), points in a uniform rectangular grid are used; for the finite-element method (b), the area is divided into triangular elements.

illustrates the discretization of the waveguide cross section and the mode field distribution in finite-difference and finite-element methods.

A relatively uncomplicated but effective finite-difference method for calculating the fundamental mode of a channel waveguide is now described in more detail. The field values at the points of a rectangular uniform grid is written as a matrix, $E(x_i, y_j) = E_{ij}$. The spacing between neighboring grid points is h. The field values at the edge points of the grid rectangle are set to 0 as a boundary condition for the guided mode—the calculation area has to be large enough for the guided modes to fulfill this condition with good accuracy. To start the calculation, a trial solution is first chosen to give initial values for the E_{ij}. This can be, for example, a Gaussian distribution concentrated at the waveguide core. The wave equation in finite difference form is

$$\frac{E_{i+1,j} + E_{i-1,j} + E_{i,j+1} + E_{i,j-1} + 2E_{ij}}{h^2} = [\beta^2 - (k_0 n_{ij})^2]E_{ij} \qquad (4.69)$$

From this the new value for E_{ij} can be solved in explicit form in terms of the values at the four neighboring points. For the value of β, a numerically integrated value is used, given by the Rayleigh quotient (4.68) and also written in a finite-difference form. This is an iterative method. Each iteration step involves first the calculation of a trial value for the propagation constant and second the calculation of the new field values from the present field values in the matrix. The calculation will converge to the fundamental mode solution.

A substantial acceleration of the iteration is achieved when the field values in the matrix are updated row by row and column by column, so that two of the neighboring field values used have already been updated at the ongoing iteration step. This is known as the *Gauss-Seidel method*.

Even faster calculation is possible when the change in field matrix values is multiplied by a suitable overrelaxation factor. The optimum overrelaxation factor for a given type of problem has to be found by trial and error. A typical starting value for the factor can be 1.1. Overrelaxation works only in connection with the Gauss-Seidel method.

The described method is quite efficient, even with a personal computer. Examples of a fundamental mode propagation constant calculated with this method are shown in Figure 4.12. The normalized parameters for diffused channel waveguides are defined as in Figure 4.10. The values obtained by the finite difference method are also compared with those given by the effective-index method. The former values agree quite well with those calculated by Lamouche and Najafi with a scalar finite element method [24].

Figure 4.13 shows some examples of guided-mode field distributions calculated with the described finite-difference method.

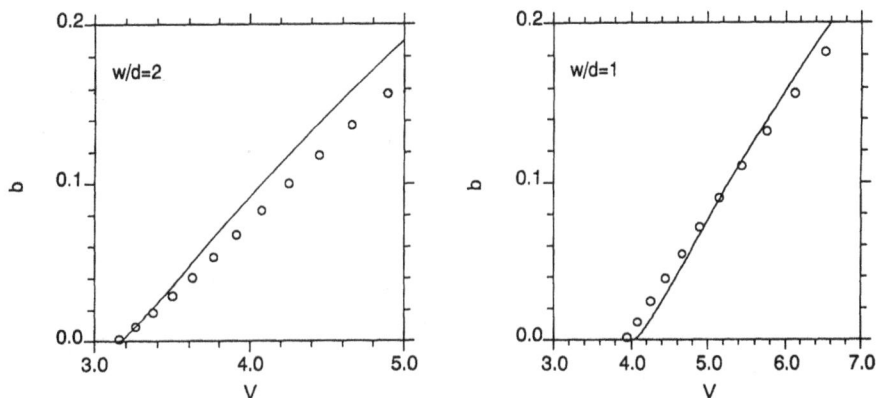

Figure 4.12 Calculated normalized propagation constant values *b* for diffused channel waveguides as functions of normalized frequency *V*. Results are shown for two ratios of mask opening width *w* to diffusion depth *d*. Values given by the scalar finite difference method are marked with circles, solid lines show the TE-mode propagation constants given by the effective index method.

4.3.5 Beam Propagation Method

A quite general problem in modeling integrated optics waveguide circuits is this. The optical field at the input plane is known; for example, the field coupled into the circuit from the input fiber. The light will propagate through the circuit more or less parallel to the axis that is perpendicular to the input plane. The problem is to calculate, from the known input field distribution and the known refractive-index distribution of the waveguide circuit, the propagation of the optical field through the circuit.

One way is to use theory of normal modes. To apply this, the circuit must be approximated by successive waveguide sections, uniform in the direction of propagation. For each section, the optical normal modes are solved and the fields at the interfaces are represented as superpositions of orthogonal mode fields. Propagation of each mode through a section can be simply calculated when the mode propagation constants are known. This method is complicated, because the continuum of radiation modes must be solved as well, and if there is much variation in the refractive-index profile along the propagation direction, enormous complications will be a result.

For this kind of modeling, a calculation algorithm for a propagation of a general field distribution through an arbitrary refractive-index distribution is needed. This chapter will briefly describe the most widely used algorithm for such

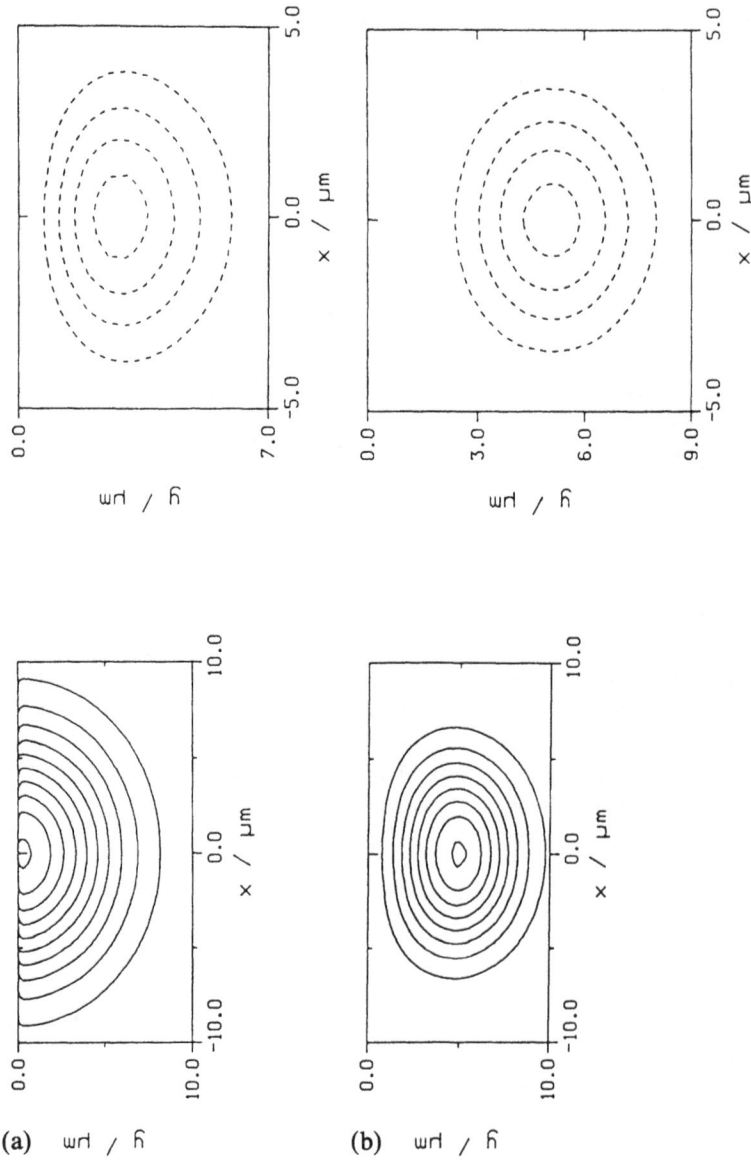

Figure 4.13 On the left, channel waveguide refractive-index distributions are shown, with contours for index increase from substrate index at intervals of 0.001. On the right are the calculated scalar mode–field intensity distributions, with contours for intensity values 0.3, 0.5, 0.7, and 0.9—the intensity maximum is normalized to unity. Optical wavelength is 1.523 μm; (a) surface waveguide with postbake diffusion; (b) waveguide buried with migration in electric field. (From [25].)

problems. The *beam propagation method* (BPM) [26–28] is a very powerful tool for simulation of integrated optical device performances.

The optical propagation problem described is solved by finding the solution for the scalar Helmholtz equation:

$$\nabla^2 E + [k_0 n(x, y, z)]^2 E = 0 \tag{4.70}$$

with the known boundary condition $E(x, y, 0)$ at the input plane $z = 0$. Here propagation is along the z-axis and the refractive index distribution $n(x, y, z)$ is known. The index distribution is presumed to be a relatively small perturbation from a uniform distribution:

$$n(x, y, z) = n_0 + \delta n(x, y, z) \tag{4.71}$$

First, the solution for a uniform refractive-index distribution $n(x, y, z) = n_0$ is examined. The solutions for the Helmholtz equation in uniform space are plane waves, so our general solution may be represented as a superposition of these:

$$E(x, y, z) = \iint_{-\infty}^{\infty} e(k_x, k_y) e^{-ik_x x} e^{-ik_y y} e^{-ik_z z} \, dk_x dk_y \tag{4.72}$$

with relation $k_z = [(n_0 k_0)^2 - k_x^2 - k_y^2]$. The plane wave amplitudes $e(k_x, k_y)$ may be obtained from the field distribution at $z = 0$ by taking Fourier transform of (4.72):

$$e(k_x, k_y) = (1/2\pi)^2 \iint_{-\infty}^{\infty} E(x, y, 0) e^{ik_x x} e^{ik_y y} \, dx \, dy = F[E(x, y, 0)] \tag{4.73}$$

where F represents the Fourier transform operation. The algorithm for calculating field at an arbitrary plane perpendicular to the z-axis, by combining (4.72) and (4.73), is

$$E(x, y, z) = F^{-1}\{F[E(x, y, 0)] e^{-ik_z z}\} \tag{4.74}$$

F^{-1} represents the inverse Fourier transform given by the integration in (4.72). This is the field-diffraction operator for propagation in uniform space.

The effect of the refractive-index perturbation for a propagation through a short distance Δz is to slightly perturb the phase of the field-front distribution. This can be represented by multiplying with a lens correction operator $\exp(-ik_0 \Delta n^2 \Delta z / 2n_0)$, where $\Delta n^2 = n(x, y, z)^2 - n_0^2$.

The BPM algorithm for a propagation through a short step is a combination of diffraction and lens correction operators:

$$E(x, y, z + \Delta z) =$$

$$F^{-1}\{e^{-ik_z\Delta z/2}F\{\exp(-ik\Delta n^2\Delta z/2n_0)F^{-1}[e^{-k_z\Delta z/2}F[E(x, y, z)]]\}\} \qquad (4.75)$$

This single formula is the essence of the BPM. The diffraction operator is split to be used both before and after the lens correction. The simulation is made by a combination of a large amount of short (on the order of a wavelength) propagation steps. Two successive diffraction operators can be combined to speed up the calculation. The field and refractive-index distribution are actually represented as discrete values in a uniform grid. The power and stability of BPM is due largely to the well-known fast Fourier transform algorithm for discrete Fourier transform calculation.

The calculation area must be selected to have enough grid points in the transverse direction to include all the guided fields. At the edges of the grid, the reflections of the radiated light may be eliminated by multiplying with a smooth absorber function at each propagation step [29].

The applicability conditions for the BPM are relatively small refractive-index variations and more or less paraxial propagation in one direction [30].

BPM is often combined with the effective-index method to make the problem two dimensional. Then only one-dimensional Fourier transform is needed, and the calculation is much faster.

REFERENCES

1. Terai, R., and R. Hayami, "Ionic Diffusion in Glasses," *J. Non-Cryst. Solids*, Vol. 18, 1975, pp. 217–264.
2. Garfinkel, H.M., "Ion-Exchange Equilibrium Between Glass and Molten Salts," *J. Phys. Chem.*, Vol. 72, 1968, pp. 4175–4181.
3. Chludzinski, P., R.V. Ramaswamy, and T.J. Anderson, "Ion Exchange Between Soda-Lime-Silica Glass and Sodium Nitrate-Silver Nitrate Molten Salts," *Phys. Chem. Glasses*, Vol. 28, 1987, pp. 169–173.
4. Stewart, G., and P.J.R. Laybourn, "Fabrication of Ion-Exchanged Optical Waveguides from Dilute Silver Nitrate Melts," *IEEE J. Quantum Electron.*, Vol. QE-14, 1978, pp. 930–934.
5. Ross, L., N. Fabricius, and H. Oeste, "Single Mode Integrated Optical Waveguides by Ion-Exchange in Glass," Proc. Fifth Annual European Fibre Optic Communications and Local Area Networks Exposition, EFOC/LAN 87, Basel, June 3–5, 1987, pp. 99–102.
6. Findakly, T., and E. Garmire, "Reduction and Control of Optical Waveguide Losses in Glass," *Appl. Phys. Lett.*, Vol. 37, 1980, pp. 855–856.
7. Najafi, S.I., P.G. Suchoski, and R.V. Ramaswamy, "Silver Film-Diffused Glass Waveguides: Diffusion Process and Optical Properties," *IEEE J. Quantum Electron.*, Vol. QE-22, 1986, pp. 2213–2218.

8. Tervonen, A., S. Honkanen, and M. Leppihalme, "Ion-Exchange Processes in Glass for Fabrication of Waveguide Couplers," *Proc. SPIE—Int. Soc. Opt. Eng.* (USA), Vol. 862, 1988, pp. 32–39.

9. Abou-El-Leil, M., and A.R. Cooper, "Analysis of Field-Assisted Binary Ion Exchange," *J. Am. Ceram. Soc.*, Vol. 62, 1979, pp. 390–395.

10. Kaneko, T., and H. Yamamoto, "On the Ionic Penetration of Silver Film into Glasses under the Electric Field," *Proc. Tenth Int. Congress on Glass*, Kyoto, Japan, 1974, pp. 8-79–8-86.

11. Ferziger, J.H., *Numerical Methods for Engineering Application*, John Wiley and Sons, New York, 1981, pp. 152–155.

12. Poszner, T., G. Schreite, and R. Muller, "Development and Characterization of Ag^+-Na^+ Exchanged Waveguides in Glass," *Proc. SPIE—Int. Soc. Opt. Eng.* (USA), Vol. 1085, 1989, pp. 413–418.

13. Tervonen, A. "A General Model for Fabrication Processes of Channel Waveguides by Ion Exchange," *J. Appl. Phys.*, Vol. 67, 1990, pp. 2746–2752.

14. Albert, J., and J.W.Y. Lit, "Full Modeling of Field-Assisted Ion Exchange for Graded Index Buried Channel Optical Waveguides," *Appl. Opt.*, Vol. 29, 1990, pp. 2798–2804.

15. Fantone, S.D., "Refractive Index and Spectral Models for Gradient-Index Materials," *Appl. Opt.*, Vol. 22, 1983, pp. 432–440.

16. Ryan-Howard, D.P., and D.T. Moore, "Model for the Chromatic Properties of Gradient-Index Glass," *Appl. Opt.*, Vol. 24, 1985, pp. 4356–4366.

17. Brandenburg, A., "Stress in Ion-Exchanged Glass Waveguides," *J. Lightwave Technol.*, Vol. LT-4, 1986, pp. 1580–1593.

18. Albert, J., and G.L. Yip, "Stress-Induced Index Change for K^+-Na^+ Ion Exchange in Glass," *Electron. Lett.*, Vol. 23, 1987, pp. 737–738.

19. Kogelnik, H., "Theory of Dielectric Waveguides," in T. Tamir (ed.), *Integrated Optics*, Springer-Verlag, Berlin, 1975, pp. 13–81.

20. Chiang, K.S., "Construction of Refractive-Index Profiles of Planar Dielectric Waveguides from the Distribution of Effective Indexes," *J. Lightwave Technol.*, Vol. LT-3, 1985, pp. 385–391.

21. Ramaswamy, V., and R.K. Lagu, "Numerical Field Solution for an Arbitrary Asymmetrical Graded-Index Planar Waveguide," *J. Lightwave Technol.*, Vol. LT-1, 1983, pp. 408–417.

22. Schweig, E., and W.B. Bridges, "Computer Analysis of Dielectric Waveguides: A Finite-Difference Method," *IEEE Trans. Microwave Theory Tech.*, Vol. MTT-32, 1984, pp. 531–541.

23. Silvester, P., "A General High-Order Finite-Element Waveguide Analysis Program," *IEEE Trans. Microwave Theory Tech.*, Vol. MTT-17, 1969, pp. 204–210.

24. Lamouche, G., and S.I. Najafi, "Accurate Analysis of Ordinary and Grating Assisted Ion-Exchanged Glass Waveguides," *Proc. SPIE—Int. Soc. Opt. Eng.* (USA), Vol. 1338, 1990, pp. 54–63.

25. Tervonen, A., P. Pöyhönen, S. Honkanen, M. Tahkokorpi, and S. Tammela, "Examination of Two-Step Fabrication Methods for Single-Mode Fiber Compatible Ion-Exchanged Glass Waveguides," *Appl. Opt.*, Vol. 30, 1991, pp. 338–343.

26. Feit, M.D., and J.A. Fleck, Jr., "Light Propagation in Graded-Index Optical Fibers," *Appl. Opt.*, Vol. 17, 1978, pp. 3990–3998.

27. Van Roey, J., J. van der Donk, and P.E. Lagasse, "Beam-Propagation Method: Analysis and Assessment," *J. Opt. Soc. Am.*, Vol. 71, 1981, pp. 803–810.

28. Lagasse, P.E., and R. Baets, "Application of Propagating Beam Methods to Electromagnetic and Acoustic Wave Propagation Problems: A Review," *Radio Sci.*, Vol. 22, 1987, pp. 1225–1233.

29. Saijonmaa, J., and D. Yevick, "Beam-Propagation Analysis of Loss in Bent Optical Waveguides and Fibers," *J. Opt. Soc. Am.*, Vol. 73, 1983, pp. 1785–1791.

30. Thylen, L., "The Beam Propagation Method: An Analysis of Its Applicability," *Opt. Quantum Electron.*, Vol. 15, 1983, pp. 433–439.

Chapter 5
Optical Waveguide-Characterization Techniques
Ming-Jun Li

Northern Telecom Canada Limited
Saskatoon, Saskatchewan

Slab and channel waveguides are the most basic elements in integrated optical circuits. An accurate knowledge of the optical characteristics of these waveguides (e.g., refractive-index profile, mode profile, cutoff wavelength, propagation losses) is necessary for device design and specification.

This chapter covers the optical waveguide-characterization techniques, which are used mostly in integrated optics. Some of the techniques have not been used for glass waveguide characterization. We explain them because we think that they can be used to test glass waveguides. In Sections 5.1–5.5, we present basic parameter measurements, such as effective refractive index, mode profile, refractive-index profile, transmission spectrum, and propagation losses. In Sections 5.6 and 5.7 we describe characterization of grating-assisted and rare-earth-doped waveguides, in which there has been an increasing interest recently.

5.1 EFFECTIVE REFRACTIVE-INDEX MEASUREMENT

The effective refractive index is a basic parameter that characterizes a guided mode. For a guided mode of order m, its effective refractive index is defined as [1]

$$N_m = \frac{\beta}{k} \tag{5.1}$$

where β is the propagation constant of the mode and $k = \lambda/2\pi$ is the free-space wavenumber.

Prism-coupling and grating-coupling techniques can be used to determine the effective refractive index of the guided modes in a waveguide.

5.1.1 Prism-Coupling Method

The principle of the prism-coupling method [2–3] is shown schematically in Figure 5.1. An incident light beam enters the prism at angle θ. At the prism base, the light beam forms an angle ϕ to the normal. This angle ϕ determines the phase velocity in the z direction of the incident beam in the prism and in the gap between the prism and the waveguide:

$$v_i = \frac{c}{n_p \sin\phi} \tag{5.2}$$

where c is the velocity of light in free space, and n_p is the refractive index of the prism. Efficient coupling of light into the waveguide occurs only when we choose the angle ϕ such that v_i is equal to the phase velocity, v_m, of one of the guided modes of the waveguide ($m = 0, 1, 2, \ldots$). Because

$$v_m = \frac{c}{N_m} \tag{5.3}$$

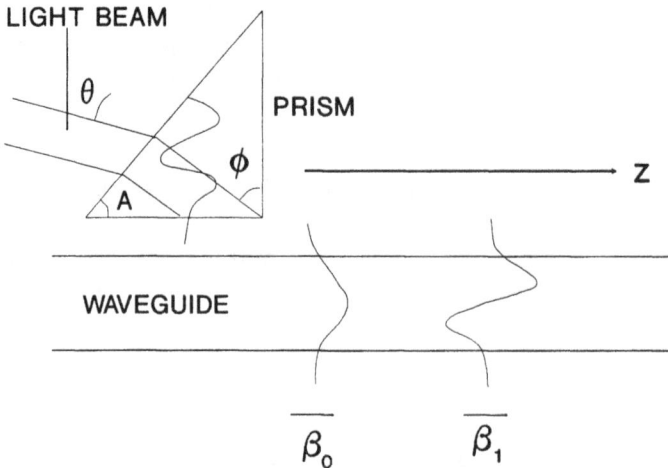

Figure 5.1 Principle of prism coupling.

the effective refractive index N_m of the mth mode is determined by

$$N_m = n_p \sin\phi_m \tag{5.4}$$

Under this condition, ϕ_m is larger than the critical angle of total internal reflection at the prism-air interface. The light is totally reflected at the base of the prism, and a stationary wave is formed in the prism. In the air gap, there exists an evanescent field, as shown in Figure 5.1. This evanescent field interacts with the evanescent field of the guided mode, resulting in coupling the incident light to the mode.

In eq. (5.4), ϕ_m cannot be measured directly. However, it is easy to relate ϕ_m to θ_m, which is directly measurable [2]. Then, N_m is given by

$$N_m = n_p \sin[\sin^{-1}\left(\frac{\sin\theta_m}{n_p}\right) + A] \tag{5.5}$$

where A is the base angle of the prism.

The experimental arrangement for measuring the coupling angles is shown schematically in Figure 5.2. The waveguide to be measured is mounted on a goniometer. Two prisms are pressed against the waveguide by means of spring-loaded clamps. Light from a laser is coupled into the waveguide through the first prism to excite the guided modes. The light in these modes is coupled out of the waveguide through the second prism. The output light forms a multiline pattern, which is called *m-lines*. Each line corresponds to a guided mode in the waveguide. The angles θ_m between these m-lines and the normal of the prism are measured by using a telescope also mounted on the goniometer.

As an example, the measured effective refractive indices of an ion-exchanged glass waveguide are given in Table 5.1. The waveguide was made by silver-ion exchange in Corning 0211 glass in pure silver nitrate at 270°C for 4.5 hrs.

The prism coupling is a very simple and fast technique, and it gives accurate results. An accuracy of about 10^{-4} can be achieved. However, buried waveguides, waveguides with high index materials, and waveguides made by soft materials are not always amenable to characterization using the prism-coupling method.

5.1.2 Grating-Coupling Method

Gratings also can be used to couple the light into or out of waveguides [1, 4]. Figure 5.3 schematically shows a grating coupler. The grating with a period Λ at the surface of the waveguide generates a small perturbation to a mode (guided or radiation mode) of the waveguide, which causes coupling of the light in the mode to another mode determined by the phase-matching condition. When phase match-

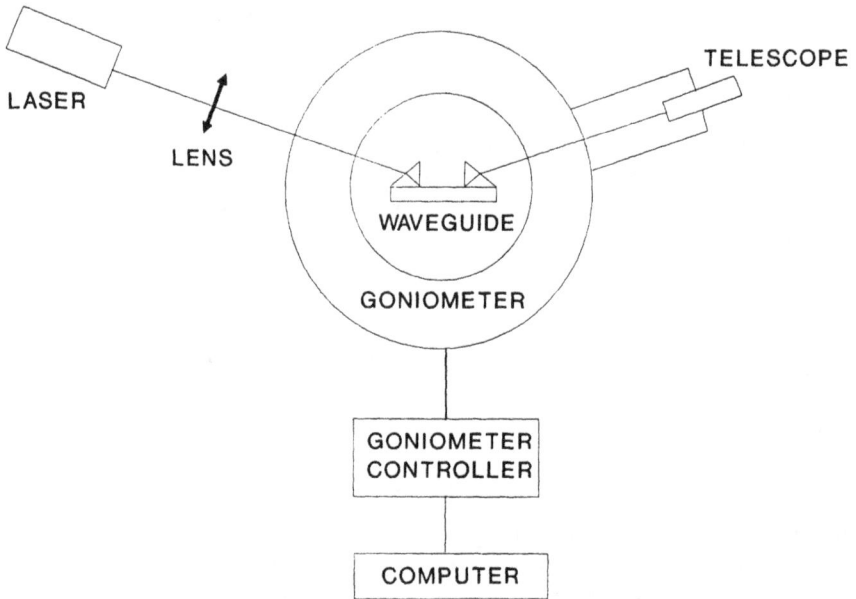

Figure 5.2 Experimental setup for measuring coupling angles in the prism-coupling method.

ing has taken place between a guided mode and a radiation mode, an efficient transfer of power between these two modes occurs. The phase-matching condition in this case can be written as

$$\beta_m = \beta_r + \frac{2j\pi}{\Lambda} \qquad j = 0, \pm 1, \pm 2, \ldots \tag{5.6}$$

where $\beta_m = \dfrac{2\pi}{\lambda} N_m$ and $\beta_r = \dfrac{2\pi}{\lambda} \sin\theta_m$ are the propagation constants of the guided and the radiation modes, respectively, and j is the grating order. Usually the first-order grating ($j = 1$) is used. In this case, the effective refractive index N_m can be expressed as

$$N_m = \sin\theta_m + \frac{\lambda}{\Lambda} \tag{5.7}$$

The grating-coupling method is suitable for measuring the effective indices for both slab and channel waveguides. The angles θ_m can be measured using the

Table 5.1
Effective Refractive Indices Measured by the Prism-Coupling Method
($A = 45°$; $\lambda = 0.6328$ μm; $n_p = 1.71653$)

Mode	θ (°)	N
TE$_0$	37.0167	1.5624
TE$_1$	34.7832	1.5482
TE$_2$	32.9500	1.5358
TE$_3$	31.4167	1.5250

same setup as described in Section 5.1.1, except the light is coupled out of the waveguide by a grating, not by a prism, and a proper coupling technique should be used to couple the light into the waveguide according to the type of the waveguide (for example, prism coupling for slab waveguides, end coupling for channel waveguides). Once θ is known, the corresponding N_m is calculated by Eq. (5.7).

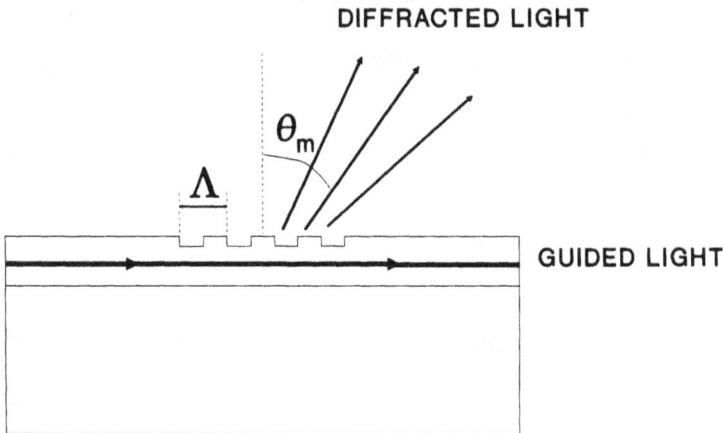

Figure 5.3 Grating coupling schematic.

Table 5.2 gives the effective refractive indices of a channel glass waveguide measured by grating coupling method [4]. The channel waveguides were made by silver-ion exchange through a 10 μm wide opening in an aluminum mask on the surface of a Corning 0211 glass substrate. The ion exchange was carried out in pure silver nitrate at 270°C for 4.5 hours. The grating was made by using a standard

Table 5.2
Effective Refractive Indices Measured by the Grating-Coupling Method
(λ = 0.6328 μm; Λ = 0.4227 μm)

Mode	θ (°)	N
TE$_0$	3.8048	1.5634
TE$_1$	2.7603	1.5452
TE$_2$	2.0148	1.5322

holographic setup and plasma etch. Grating coupling is also an accurate method. The accuracy is comparable to that of prism coupling method, about 10^{-4}.

5.2 MODE-PROFILE CHARACTERIZATION

The mode profile, or more specifically the intensity distribution of a mode, is an important characteristic of a waveguide. Many properties of the waveguide, such as its coupling efficiency to an optical fiber or another waveguide, and its interaction with a grating, are related to the mode profile. Through mode-profile characterizations, we can get some information about, for example, the number of modes, as well as their dimensions and the symmetry of the waveguide. As we will see in Section 5.3.3, the mode profile can also be used to determine the refractive-index profile of the waveguide.

The experimental apparatus for mode profile characterization is shown in Figure 5.4. Light from a laser is end coupled into a waveguide to excite a mode

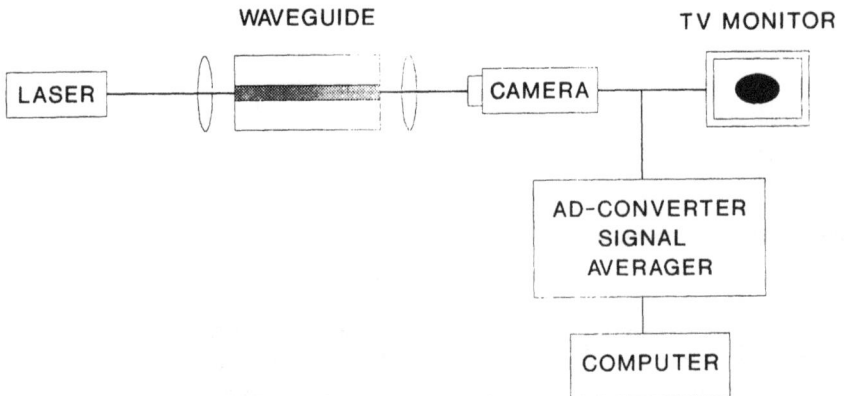

Figure 5.4 Experimental apparatus for measuring mode profiles.

(for a slab waveguide, the prism-coupling technique can also be used). Because the mode-field dimensions are in the micrometer range, we have to magnify the near-field pattern by using a microscope objective. The magnification of the system can be determined by measuring a specimen whose dimensions are known. The image is focused onto a video camera and then converted to digital data that can be processed by a computer.

To get correct mode profiles, precautions must be taken. First, the focusing must be extremely accurate to reduce the error in magnification. Second, the camera should have a very good linearity between optical power and digital output. If the camera is not linear, the result must be corrected by the response of the camera.

Figure 5.5 is a photograph of the fundamental mode and its transversal and lateral profiles in a potassium and silver double-ion-exchanged channel waveguide [5]. The waveguide was made first by potassium-ion exchange in pure potassium

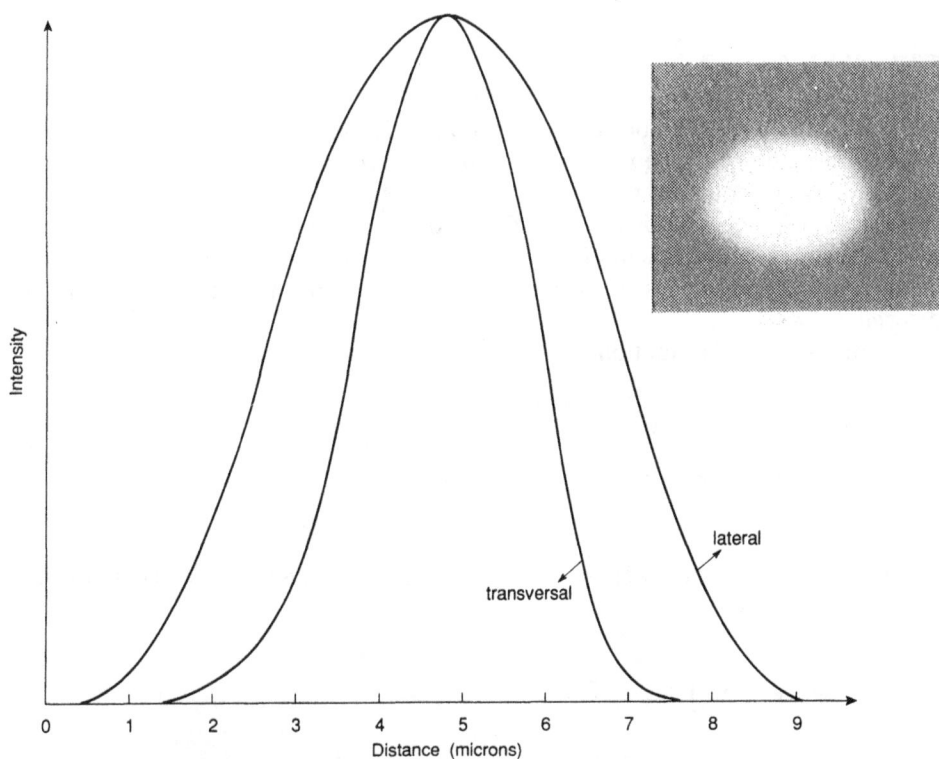

Figure 5.5 Photograph of the fundamental mode and its profiles at 1.3 μm of a potassium and silver double-ion-exchanged waveguide.

nitrite at 400°C for 140 min and then by silver-ion exchange in pure silver nitrite at 300°C for 300 min.

5.3 INDEX-PROFILE DETERMINATION

The properties of an optical waveguide are determined mainly by its refractive-index profile. The index-profile characterization thus has great importance in integrated optics.

Different techniques can be employed to determine the index profiles in slab and channel waveguides. Simple methods such as WKB [6–7] and inverse WKB [8–10] can be used to obtain the index profiles of slab waveguides. Methods for determining index profiles of channel waveguides are much more complicated. In this section we will discuss the mode near-field method [11–13] to reconstruct the index profiles of these waveguides.

5.3.1 WKB Method

WKB method, usually associated with Wentzel, Kramers, Brillouin, and Jeffreys, is sometimes known as the *phase-integral method*. It is an approximate method for solving the wave equation for slab waveguides [6]. If the refractive-index function of a slab waveguide is known, the WKB method can be used to find the effective refractive index and the field distribution of all modes. On the other hand, if the effective refractive indices of the waveguide are known, the index profile can be determined.

For an index distribution

$$
n(x) = \begin{cases} n_c & x < 0 \\ n(x) & x > 0 \end{cases}
\tag{5.8}
$$

and for $n(x) \gg n_c$ and $n(0) \approx n(x)$, WKB approximation yields the following eigenvalue equation:

$$
\int_0^{x_m} [n^2(x) - N_m^2]^{1/2} dx = \frac{4m + 3}{8} \qquad m = 0, 1, 2, \ldots, M - 1
\tag{5.9}
$$

where x_m is defined by $N_m = n(x_m)$, and M is the number of modes.

If $n(x)$ is a function with m parameters, that is,

$$n(x) = f(p_1, p_2, p_3, \ldots, p_m, x) \tag{5.10}$$

for a waveguide that supports m modes or more, the parameters $p_1, p_2, p_3, \ldots, p_m$ can be determined by fitting the measured effective refractive indices into eq. (5.9).

Let us take an example to see how to determine the refractive-index profile of ion-exchanged waveguides by WKB method. For waveguides made by ion exchange, the index profile can be described by the complementary error function [14]:

$$n(x) = \Delta n \operatorname{erfc}(x/d) + n_s \tag{5.11}$$

with $d = 2\sqrt{Dt}$ defined as the waveguide depth, D is the diffusion coefficient, t is the diffusion time, n_s is the refractive index of the substrate, $\Delta n = n(0) - n_s$ is the maximum index change, and

$$\operatorname{erfc}(x) = \frac{2}{\sqrt{\pi}} \int_x^\infty \exp(-\alpha^2) \, d\alpha \tag{5.12}$$

The refractive-index profile of ion-exchanged waveguides is thus characterized by Δn and d. The analysis of these waveguides can be generalized by using the normalized frequency v and the normalized propagation constant b, which are defined as [7]

$$b = \frac{N_m^2 - n_s^2}{2n_s \Delta n} \tag{5.13}$$

$$v = kd(2n_s \, \Delta n)^{1/2} \tag{5.14}$$

where k is the free-space wavenumber. Examples of b-v curves calculated with the WKB method are given in Chapter 2 (see Figure 2.12). The two parameters Δn and d for a particular waveguide are found by fitting measured effective indices (as explained in Section 5.1) to these curves.

In WKB method, we have to assume an index profile function characterized by a certain number of parameters. If there are only few parameters, the calculation is not difficult. However, for some waveguides, it is not easy to describe their index profiles with simple functions. It is evident that, with a complicated function which has many parameters, this method becomes very computer-time absorbing. In this case, we can use the inverse WKB method to find the index profiles.

5.3.2 Inverse WKB Method

This method was first proposed by White and Heidrich [8], then improved by Chiang [9] and Hertel and Menzler [10]. The basic equation is eq. (5.9). The idea is to construct the index profile from the measured effective indices.

We use White and Heidrich's approach first to show the principle of this method. To determine $n(x)$, we proceed by writing eq. (5.9) as a sum of integrals:

$$\sum_{k=1}^{i} \int_{x_{k-1}}^{x_k} [n^2(x) - n_i^2]^{1/2} \, dx = \frac{4m_i + 3}{8} \tag{5.15}$$

In writing (5.15), we have used n_i and m_i instead of N_m and m to generalize the results so that they can be applied to Chiang's approach as well. For White and Heidrich's approach, which we are discussing now,

$$n_0 = n(0), \qquad n_i = N_{i-1}, \qquad m_i = i - 1, \qquad i = 1, 2, \ldots, M$$

Next we assume that $n(x)$ is a piecewise linear function connecting the measured values of n_i; that is,

$$n(x) \approx n_k + \frac{n_{k-1} - n_k}{x_k - x_{k-1}}(x_k - x) \qquad x_{k-1} \leq x \leq x_k \tag{5.16}$$

If we let $n(x) + n_i$ be replaced by a midpoint value, that is,

$$n(x) + n_i \approx \frac{n_{k-1} + n_k}{2} + n_i \qquad x_{k-1} \leq x \leq x_k \tag{5.17}$$

the solution for x_i would be

$$x_i = x_{i-1} + \left[\left(\frac{3}{2}\right) \left(\frac{n_{i-1} + 3n_i}{2}\right)^{-1/2} (n_{i-1} - n_i)^{-1/2} \right]$$

$$\cdot \left\{ \left(\frac{4m_i + 3}{8}\right) - \frac{2}{3} \times \sum_{k=1}^{i-1} \left(\frac{n_{k-1} + n_k}{2} + n_i\right)^{1/2} \left(\frac{x_k - x_{k-1}}{n_{k-1} - n_k}\right) \right.$$

$$\left. \cdot [(n_{k-1} - n_i)^{3/2} - (n_k - n_i)^{3/2}] \right\} \qquad \text{for} \quad i = 2, 3, \ldots, M \tag{5.18}$$

$$x_1 = \frac{9}{16} \left(\frac{n_0 + 3n_1}{2} \right)^{-1/2} (n_0 - n_1)^{-1/2} \tag{5.19}$$

Therefore, if the surface index n_0 is known, x_i can be calculated by using eqs. (5.18) and (5.19), and $n(x)$ is determined approximately by eq. (5.15). However, usually we measure only the effective indices N_0, N_1, . . . , N_{M-1} (see Section 5.1), the surface index n_0 is unknown. Because normally the index profile is a smooth curve, n_0 can be chosen as the value that gives the smoothest profile.

Figure 5.6 shows an index profile of a potassium-ion-exchanged waveguide calculated by the inverse WKB method using the measured effective indices. The waveguide was fabricated in a Corning 0211 glass substrate by ion exchange in pure potassium nitrate at 400°C for 2 hrs.

The approach of White and Heidrich is very simple. However, their method is accurate only for highly multimode waveguides, as the number of straight line segments that construct the profile is equal to the number of guided modes.

Chiang [9] improved the inverse WKB method by constructing an effective index function $N(m)$. The effective index function is constructed by interpolating the measured effective indices N_0, N_1, . . . , N_{M-1} by using the Gregory-Newton interpolation formula, in which N_0 is used as the baseline and forward finite difference is employed. $N(m)$ constructed in this way is in general a polynomial of order $M - 1$. With $N(m)$, we can obtain the surface index n_0 from eq. (5.9). By putting $x_m = 0$ in eq. (5.9), we get $m = -0.75$. Therefore, the surface index is

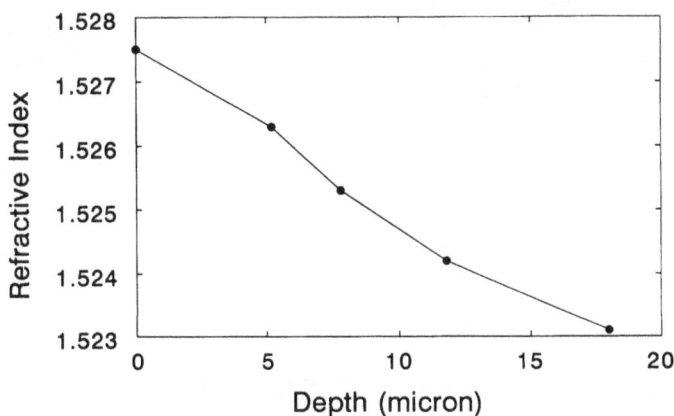

Figure 5.6 Refractive index profile of a potassium-ion-exchanged waveguide reconstructed by the inverse WKB method.

determined by

$$n_0 = N(-0.75) \tag{5.20}$$

Once the effective-index function is available, we can use the inverse WKB method to calculate the index profile. The equations used to calculate the index profile are the same as eqs. (5.15)–(5.19), but the effective indices n_i ($i = 1, 2, \ldots$) and the corresponding m_i ($n_i = N(m_i)$) that appear in the equations are sampled from the effective-index function instead of the measured effective indices. Now m_i is not necessarily an integer. The profile constructed in this approach can be as smooth as desired depending on number of samples chosen. Profile construction from five or more modes gives very satisfactory results.

A further improvement was made by Hertel and Menzler [10]. In their method, the substrate index n_s is assumed to be known, and the behavior of index profile near the waveguide surface and at great depth is imposed. By properly enforcing the profile form near the surface and at great depth, very accurate results can be obtained. Well-behaved profiles such as Gaussian or exponential can be reconstructed within experimental accuracy from three modes, and even two modes may provide useful information.

5.3.3 Mode Near-Field Method

The WKB and inverse WKB methods are suitable only for slab multimode waveguides. For channel waveguides, one method for the reconstruction of index profile is the near-field method [11–13].

This method is based on the relation between the refractive index profile and the field distribution of a guided mode. It is well known that the electric component $E(x, y)$ for a guided mode satisfies the following scalar equation:

$$\nabla_t^2 E(x, y) + [k^2 n^2(x, y) - \beta^2] E(x, y) = 0 \tag{5.21}$$

with

$$\nabla_t^2 = \frac{\partial^2}{\partial x^2} + \frac{\partial^2}{\partial y^2} \tag{5.22}$$

From this equation, we can derive easily the expression for the refractive index profile:

$$n^2(x, y) = \frac{\beta^2}{k^2} - \frac{1}{k^2 E} \nabla_t^2 E \tag{5.23}$$

The near-field intensity $P(x, y)$ that is directly measurable is proportional to the square of the field component $E(x, y)$. The index profile is therefore expressed as a function of $P(x, y)$:

$$n^2(x, y) = \frac{\beta^2}{k^2} - \frac{1}{2k^2 P}\left[\nabla_t^2 P - \frac{1}{2P}(\nabla_t P)^2\right] \tag{5.24}$$

The first term on the right-hand side of eq. (5.24) is an unknown constant, but the remaining terms are known functions. If the refractive index is known at a certain point (for example, in the substrate), the propagation constant β can be calculated, so that the index profile can be determined everywhere.

The near-field intensity is measured with a video camera as explained in Section 5.2. The mode is excited at the single-mode operation wavelength. Because the second derivatives have to be performed numerically on the measured points, very accurate experimental data are required and smoothing procedures of these data are necessary.

5.4 TRANSMISSION SPECTRUM MEASUREMENT

The transmission spectrum of a waveguide characterizes its optical behavior as a function of wavelength. From the transmission spectrum, we can determine the cutoff wavelength of each mode, the single-mode operation region and the relative loss change with wavelength.

The experimental setup to measure the transmission spectrum is illustrated in Figure 5.7. Light from a white light source is coupled into the waveguide through a microscope objective to excite all the modes. The output light from the waveguide

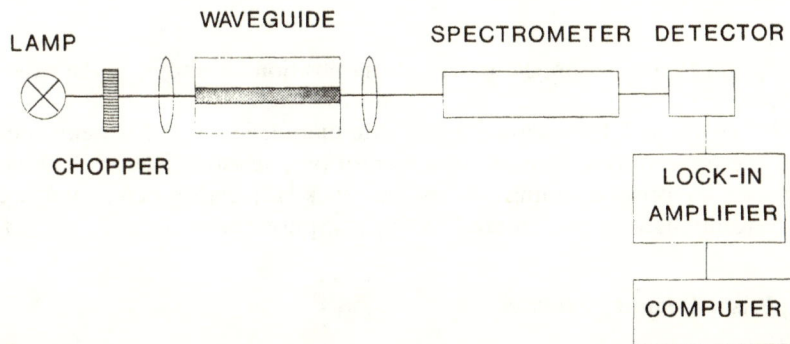

Figure 5.7 Experiment setup for measuring transmission spectra of waveguides.

is directed into a high-resolution spectrometer. The spectrometer is scanned in the desired wavelength region. At the output of the spectrometer, the light is detected and the signal is sent to a lock-in amplifier. The amplified signal data is then stored in a computer.

Examples of typical transmission spectra of potassium- and silver-ion-exchanged waveguides are given in Chapter 6. From a transmission spectrum, we can get the following information:

- Cutoff wavelength of each mode and cutoff wavelength of the waveguide. There are a number of steps in a transmission spectrum (see Figures 6.1 and 6.2). Each step corresponds to a guided mode region. When a step vanishes, the corresponding mode is in cutoff. When the last step vanishes, there is no guided mode in the waveguide, the waveguide is in cutoff.
- Single-mode operation wavelength region. The wavelength region in the last step is defined as the single-mode operation region in which only the fundamental mode propagates.
- Relative propagation losses. Theoretically, the transmitted intensity should decrease with the cutoff of each mode. Therefore, a lower transmitted intensity at short wavelengths indicates that the losses of the waveguide are higher at shorter wavelengths than at longer wavelengths.

5.5 LOSS MEASUREMENT

Its propagation losses characterize the quality of a waveguide. The losses can be caused by absorption, scattering, radiation, nonlinear effects, bends, and so forth. We use the attenuation coefficient, α, to describe the losses in a waveguide; this is defined by

$$\alpha = \frac{10 \log(P_0/P_1)}{z_1 - z_0} \text{ (dB/cm)} \tag{5.25}$$

where P_0 and P_1 are the optical power at the positions z_0 and z_1 in the waveguide, respectively.

The techniques for measuring the attenuation involve the measurement of transmitted or scattering light as a function of propagation distance. The methods used mostly are prism coupling [14–16], cut-back [17], Fabry-Perot interferometer [18], scattering light measurement [19–20] and photothermal deflection [21].

5.5.1 Prism-Coupling Method

According to eq. (5.25), to determine the propagation losses of a waveguide, we need to measure the transmitted power as a function of propagation distance in the waveguide. For slab waveguides, this measurement can be achieved by the prism-coupling method [14]. As shown in Figure 5.8, light is coupled into the waveguide by prism 1. To measure the power change as a function of propagation distance, the output prism (prism 2 in Figure 5.8) is placed at different positions along the waveguide. In this technique, it is crucial that all of the light in the guide be coupled out by the prism. To ensure a 100% coupling efficiency, Weber, Dunn, and Leibolt [14] use a sliding-output prism with an index matching oil between the prism and the waveguide. Because matching liquids are used in this method, it follows that the measurement cannot be made in this fashion for waveguides with effective index values higher than about 2.0. Furthermore, when several modes are present, the output light of individual modes tends to overlap, which makes individual mode-loss measurements difficult. When no index-matching liquid is used, the difficulty is that the clamping of the prism has to be tight enough to ensure 100% output coupling without disturbing the input coupling.

To overcome this difficulty, Won, Jaussaud, and Chartier [15] proposed a three-prism method. In this method, the measurement is independent of input and output coupling coefficients. As shown in Figure 5.9, in this technique, light is launched into the waveguide through prism 1. First, the output powers P_2 and P_3 of prisms 2 and 3 at positions of z and z_0 are measured:

$$P_2 = \gamma_2 P(z) \tag{5.26}$$

$$P_3 = \gamma_3 [P(z) - P_2] \exp[-\alpha(z_0 - z)] \tag{5.27}$$

where γ_2 and γ_3 are output coupling constants for prisms 2 and 3, $P(z)$ is the

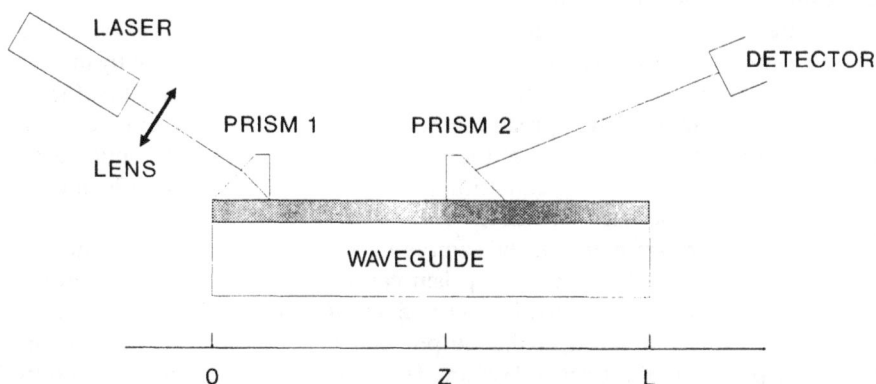

Figure 5.8 Schematic of two-prism loss measurement.

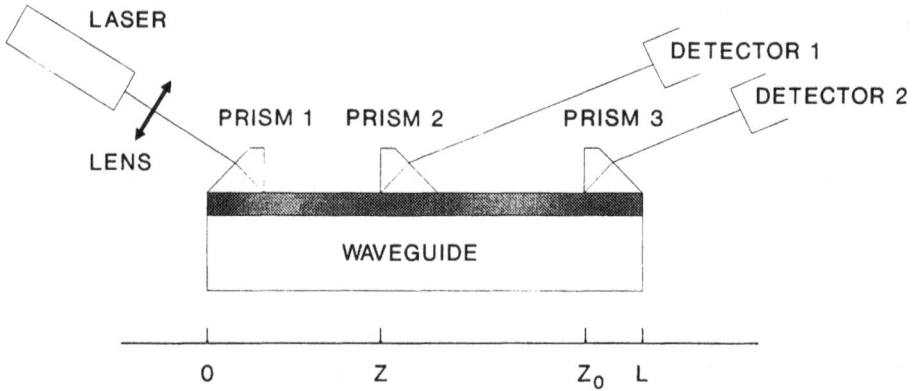

Figure 5.9 Schematic of three-prism loss measurement.

intensity of light in the waveguide at z. Then, we make $\gamma_2 = 0$ by disconnecting prism 2 from the waveguide, and measure the output power P_3^o of prism 3:

$$P_3^o = \gamma_3 P(z) \exp[-\alpha(z_0 - z)] \tag{5.28}$$

Eliminating γ_3, we find

$$P(z) = \frac{P_2 P_3^o}{P_3^o - P_3} \tag{5.29}$$

This result is independent of the coupling coefficients γ_2 and γ_3.

Figure 5.10 shows the measured results of a single-mode glass waveguide at $\lambda = 514.5$ nm for different values of γ_2 [15]. The waveguide was made by immersing a glass substrate in a KNO_3 molten bath at 368°C for 2 hrs. Figure 5.10(a) ($\gamma_2 = 100\%$) is simply the usual two-prism method without matching liquid. As it can be seen in Figure 5.10, the accuracy of the method depends on γ_2. A simple accuracy analysis based on eqs. (5.26) and (5.27) shows that a coupling efficiency γ_2 of roughly 50% gives the best accuracy.

In the three-prism method, the sample must be long enough to contain three prisms. For short samples, the third prism can be replaced by a polished output end face of waveguide, as shown in Figure 5.11 [16]. In this configuration, P_3 and P_3^o are the powers measured at the output end face of the waveguide with and without the presence of the second prism. The formula for calculating $P(z)$ remains the same as in the three-prism method (eq. (5.29)).

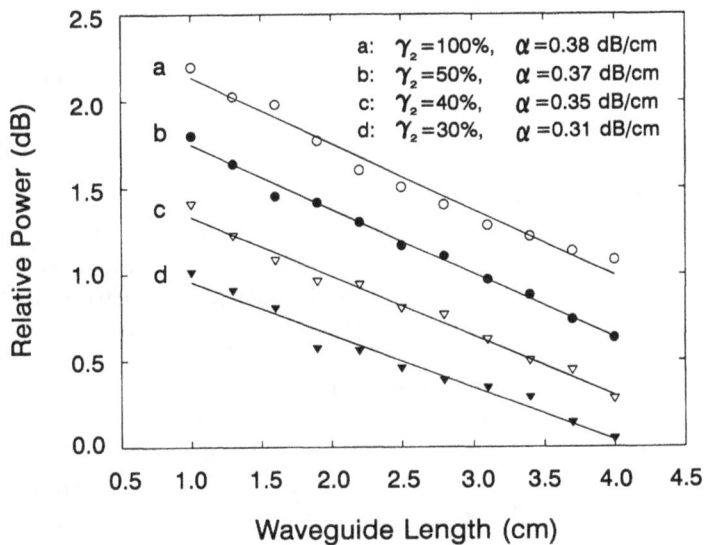

Figure 5.10 Attenuation of a single-mode waveguide measured by the three-prism method for different values of the coupling coefficient γ_2 of prism 2. (From [15].)

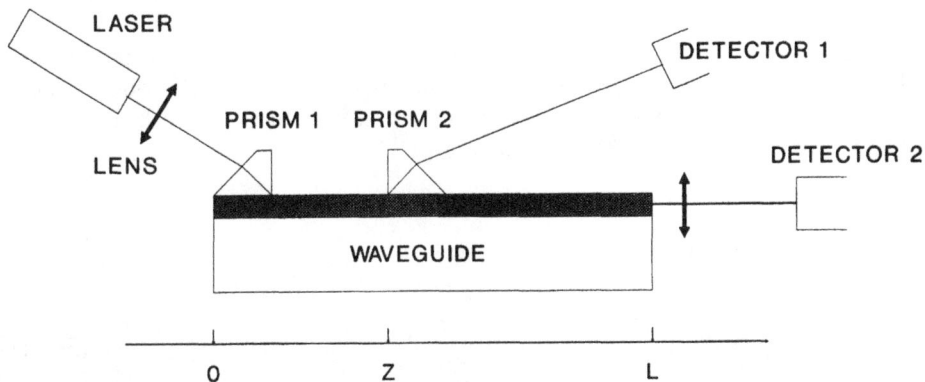

Figure 5.11 Schematic of loss measurement using two-prisms and a polished end face.

5.5.2 Cut-Back Method

Because it is difficult to couple light into and out a channel waveguide by a prism, the prism-coupling method is not convenient for measuring losses in channel waveguides. Instead, the cut-back method [17] can be used. This method is originally from the attenuation measurement of an optical fiber, in which the transmitted power is measured as a function of fiber length, by cutting the fiber without changing the launching conditions. In integrated optics, a waveguide is cut into several samples with different lengths and polished before the measurement, because it is impossible to cut and polish the waveguide during the measurement. Light is launched into the waveguides by end coupling. The transmitted powers of the samples are then measured.

To determine the attenuation coefficient, the common logarithm of the measured transmitted power is plotted as a function of sample length. A straight line is fitted with the measured data. Its slope gives the attenuation coefficient.

To obtain good results by this method, it is important to maintain the same coupling efficiency for all the samples. This affects the reproducibility of the method. We can increase the accuracy by taking several measurements of each sample. The results with similar coupling efficiency should be around a straight line. The results that are far from this line must be excluded.

5.5.3 Fabry-Perot Interferometer Method

The two end faces of a channel waveguide can form a Fabry-Perot interferometer if the reflectivity of these end faces is high enough. The loss measurement using Fabry-Perot interferometery [18] is based on the transmission of a Fabry-Perot interferometer depending on the loss of the medium inside the interferometer cavity.

The optical transmission of a Fabry-Perot interferometer can be derived from the convergent geometric series obtained by adding wave amplitudes due to successive reflections. Assuming perfectly coherent, monochromatic light, the resonant (T_R) and antiresonant (T_A) transmissions are [22]

$$T_R = P_0 \left[\frac{\gamma}{1 - \gamma^2 R} \right]^2 \tag{5.30}$$

$$T_A = P_0 \left[\frac{\gamma}{1 + \gamma^2 R} \right]^2 \tag{5.31}$$

where P_0 is the intensity coupled into the waveguide at the input, R is the geometric mean of the reflection coefficients of the two end faces, and γ is the single-pass, wave-amplitude reduction factor.

If R is known, but not P_0, the attenuation coefficient can be obtained from the ratio $K = T_R/T_A$:

$$\gamma = \frac{-20 \log\gamma}{L} = \frac{-10}{L} \log\left[\frac{1}{R} \sqrt{\frac{K-1}{K+1}}\right] \tag{5.32}$$

where L is the waveguide length. Similarly, if P_0 is known, but not R, the attenuation coefficient may be derived from the single-pass transmission (T):

$$T = P_0\gamma^2 = 4\left[\frac{1}{\sqrt{T_A}} + \frac{1}{\sqrt{T_R}}\right]^{-2}$$

$$\alpha = \frac{10}{L} \log(T/P_0) \tag{5.33}$$

If both R and P_0 are unknown, it is necessary to measure two waveguides with the same R but different lengths. By eliminating P_0 and R, we get

$$\alpha = \frac{10}{L_1 - L_2} \log\left[\sqrt{\frac{K_1 + 1}{K_1 - 1} \frac{K_2 - 1}{K_2 + 1}}\right] \tag{5.34}$$

where L_1 and L_2 are the lengths, and K_1 and K_2 are the T_R/T_A ratios, of the two samples, respectively. Note that it is not necessary to maintain the same coupling efficiency in the measurement of two samples, which makes the measurement easier.

To measure T_R and T_A, two or more oscillation cycles are needed. This implies that the refractive index, or the length of waveguide, or the wavelength used must be changed slightly. The changes can be achieved by heating the sample or by using tunable sources.

This method is very accurate for waveguides with losses lower than 0.2 dB; in fact the accuracy increases as the losses decrease. It is suitable to measure the losses of waveguides with a high refractive index, such as waveguides in semiconductors and in lithium niobate. For glass waveguides, the measurement may be more difficult to make due to the low reflectivity of the end faces. However, the problem can be solved by coating the end faces with high-reflection films.

5.5.4 Scattered-Light Measurement Method

Scattered-light measurement [19–20] is a useful technique, as it allows us to determine nondestructively the propagation losses of a waveguide and observe light propagation in devices such as directional couplers, Y-branches, modulators, switches, and deflectors.

In an optical waveguide, a guided mode continuously loses a small part of its power by Rayleigh scattering. This scattered power is proportional to the total guided power. Thus the scattered-light decrease shows the propagation losses of the guided mode.

The experimental setup to observe the scattered light is shown in Figure 5.12. It consists of a video camera (other detection systems can also be used), a video monitor, a video camera controller with an AD converter, and a computer. The principle of the measurement is simple: the light streak from the optical waveguide is detected by the video camera, and the light intensity profile is depicted on the X-Y plotter. The computer is used to command the camera and calculate the losses.

The scattered light decreases exponentially along the waveguides. By fitting the measured scattered power to a decreasing exponential function, the attenuation coefficient is obtained. The accuracy depends on the sensitivity of the detection system.

Figure 5.13 gives the results of scattered-light measurement in a single-mode glass waveguide [19]. The waveguide was made by potassium-ion exchange at 370°C for 0.5 hr through a slit of 4 μm wide and 20 mm long in an aluminum film on a glass substrate. Figure 5.13(a) shows the scattered optical field distribution along

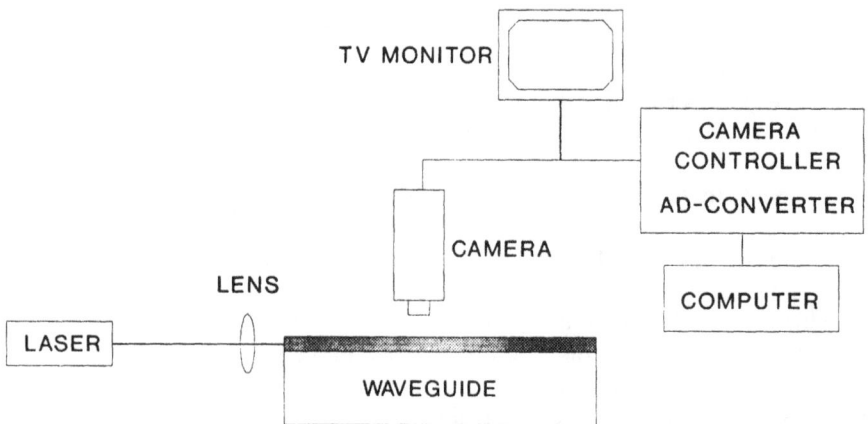

Figure 5.12 Experimental setup for scattered light measurement.

(a)

(b)

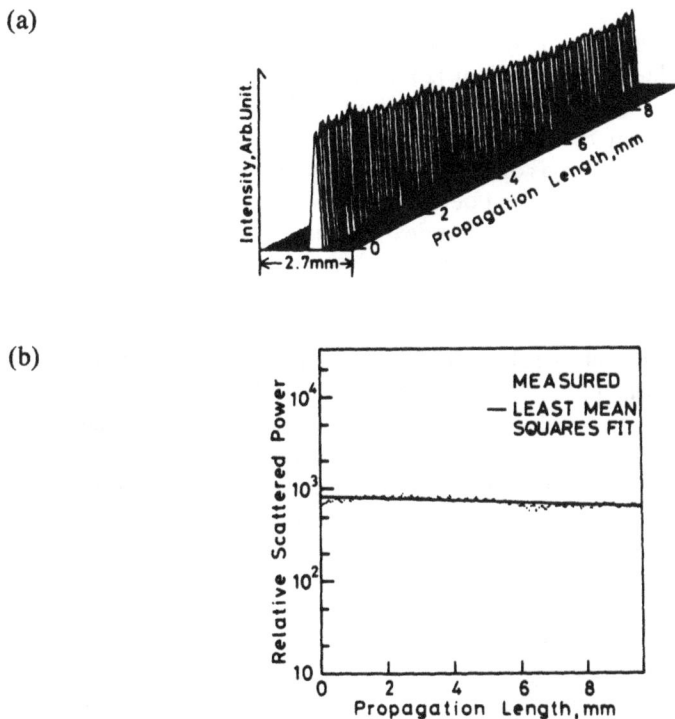

Figure 5.13 Scattered-field distribution along a single-mode channel glass waveguide (a) and peak light intensity as a function of propagation length (b). (From [19].)

the waveguide for the TE mode. The peak intensity along the propagation direction is plotted in Figure 5.13(b). A least-mean-squares fit of the measured data to a decreasing exponential function (solid line) yields propagation losses of 1.2 dB/cm. The limits of this method are that it is difficult to measure the weakly scattering waveguides, and it suffers from the highly variant scattering efficiency of random inhomogeneities in the waveguide.

5.5.5 Photothermal Deflection Method

Photothermal deflection [21] is another nondestructive method to measure the propagation losses. It is based on the photothermal deflection effect. When energy is absorbed from a beam of light, the pump beam, it produces a thermal gradient, which in turn produces a refractive-index gradient in the absorbing and surrounding media. *Photothermal deflection* (PTD) involves the refraction of a second beam of light, the probe beam, due to the gradient [23–24].

The experimental configuration of this method is shown in Figure 5.14. The pump beam is a guided light that is modulated with a chopper and end coupled into a waveguide. The probe beam is directed normally to the substrate and focused to a spot on the surface. A silicon bicell detector below the waveguide detects the probe-beam deflection due to the induced index gradient. The differential voltage of the detector is amplified and separated from noise by a lock-in amplifier.

Figure 5.15 shows the measured peak PTD signal as a function of position down the length of a glass waveguide [21]. The waveguide was made by potassium-ion exchange in a soda lime glass substrate. Masked by aluminum with a 10 μm gap width, the substrate was immersed in a KNO_3 bath at 400°C for 5.5 hrs. The PTD signal exhibits an exponential decay, the slope of a semilog graph yields the attenuation coefficient of 1.2 dB/cm, as fit by linear regression.

The advantage of this method is that scattering centers and unguided background light do not affect the measurement directly. So this method is more accurate

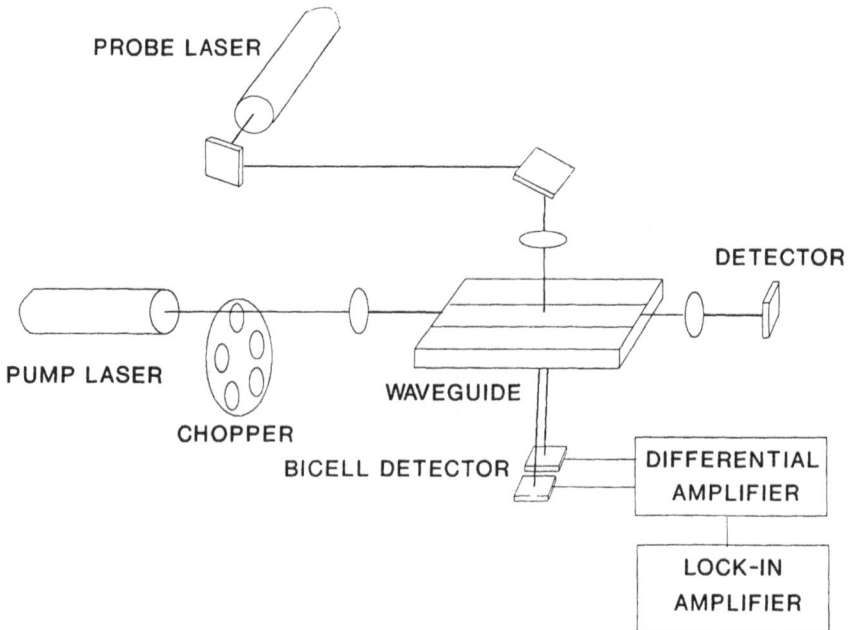

Figure 5.14 Experimental configuration of PTD method.

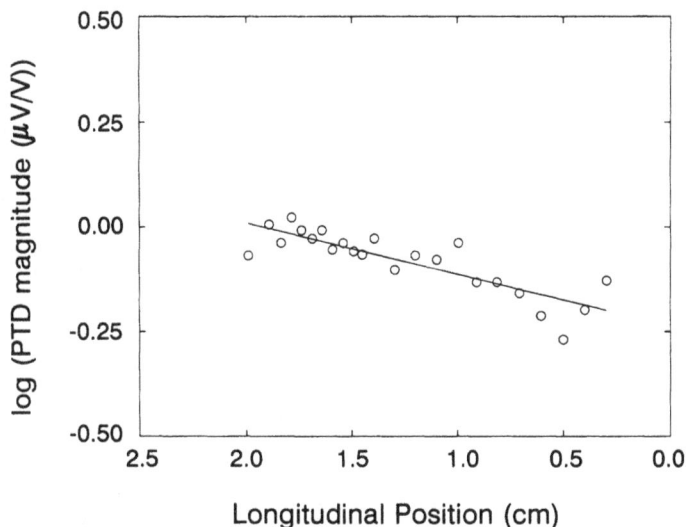

Figure 5.15 Logarithm of peak PTD magnitude as a function of distance along a glass waveguide. (From [21].)

than the scattered-light measurement. An accuracy of 0.03 dB/cm has been demonstrated using this technique.

The crossed-beam configuration of the PTD method shown previously is not suitable to substrates that are not transparent to the probe beam wavelength or have an unpolished face. For these substrates, other configurations such as reflected probe beam from the surface and probing at grazing incidence can be used.

5.6 GRATING-ASSISTED WAVEGUIDE CHARACTERIZATIONS

Grating-assisted optical waveguides are interesting components in integrated optics. They are wavelength-selective components that can be used as beam deflectors or reflectors. This section presents the basic characterizations of these waveguides.

5.6.1 Grating Period Measurement

The period of a grating determines the wavelength dependence of its performance; that is, the deflection angle for a given wavelength and the reflection peak wavelength. The principle of measuring a grating period is illustrated in Figure 5.16. For a laser beam of wavelength λ that is incident to the grating surface with an

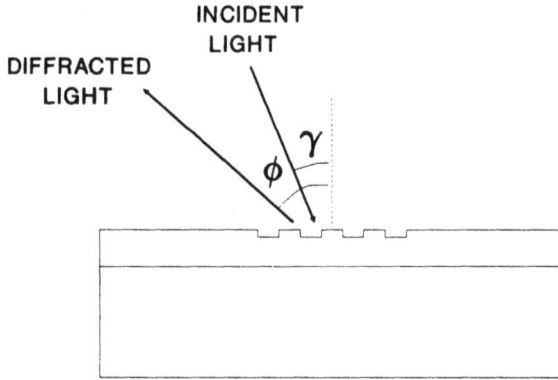

Figure 5.16 Schematic of light diffraction by a grating.

angle γ to the normal, diffracted beams are generated by the grating with angles ϕ to the normal. According to the diffraction theory, α and ϕ satisfy the following equation [25]:

$$\Lambda (\sin\gamma + \sin\phi) = \nu\lambda \qquad \nu = 0, 1, 2, \ldots \qquad (5.35)$$

where Λ is the period of the grating, ν is the diffraction order. Normally, the first-order diffraction is measured. In this case, the grating period is determined by

$$\Lambda = \frac{\lambda}{\sin\gamma + \sin\phi} \qquad (5.36)$$

In grating period measurement, the sample is mounted in a goniometer. The experiment can be simplified if we make γ equal to ϕ. To do so, we first turn the sample to the position where its surface is perpendicular to the laser beam. Then we rotate the sample so that the diffraction beam is in the same direction of the laser beam and measure the angle. In this case, the grating period is calculated by

$$\Lambda = \frac{\lambda}{2 \sin\gamma} \qquad (5.37)$$

5.6.2 SEM Analysis

The *scanning electrical microscope* (SEM) analysis is a visual method to characterize gratings. From a SEM picture, we can get information such as the period, the depth, the form, and the uniformity of a grating.

As we will see in Chapter 6, gratings in glass waveguides are made mainly by two techniques: plasma etch and ion diffusion. To analyze gratings using SEM, samples are coated with gold paladium. Secondary and backscattered electrons are used to observe gratings. The yield of the secondary electrons depend on the topography of the sample. These electrons are used to analyze etched gratings. For diffused gratings, backscattered electrons are employed, because the yield of these electrons is a function of sample composition. However, the resolution of the measurement with backscattered electrons is lower than that of the secondary electrons. Typical SEM pictures of an etched and a diffused grating in glass substrate are shown in Figure 6.10.

5.6.3   Grating-Efficiency Measurement

The interactions between a guided light and a grating can be divided into two types. The light can be diffracted or reflected by the grating depending on grating period and propagation constant of the mode.

5.6.3.1   *Diffraction-Efficiency Measurement*

The experimental setup to measure the diffraction efficiency is illustrated in Figure 5.17. Light from a laser is coupled into the waveguide, and the diffracted and transmitted powers are measured. The ratio of the two powers gives the diffraction efficiency (see Table 6.1 of Chapter 6).

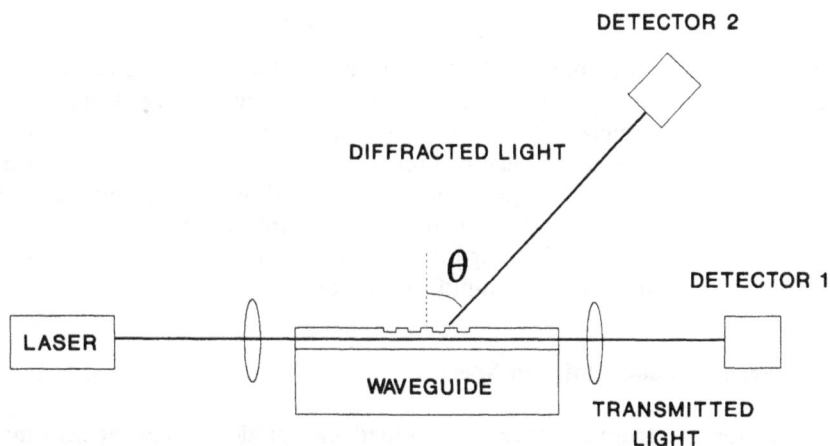

Figure 5.17 Experimental setup for measuring diffracted light by grating-assisted waveguides.

5.6.3.2 Reflection Efficiency

The light reflected by a grating is rather difficult to measure. Instead, we measure the transmitted light. The light source can be white sources, LEDs, or tunable lasers. High resolution is difficult to achieve using a white source due to the low light intensity that can be coupled into the waveguide. Good results can be obtained using LEDs or tunable lasers. The experimental setup is the same as in the measurement of the transmission spectrum explained in Section 5.4. A measured spectrum of a potassium and silver double-ion-exchanged waveguide with an etched grating is given in Figure 6.12.

5.7 RARE-EARTH-DOPED-WAVEGUIDE CHARACTERIZATION

Rare-earth-doped waveguides are useful components to make active devices such as lasers, amplifiers, and modulators. The most important property of rare-earth-doped waveguides is that they can amplify optical signals. For an optical wave of frequency ν with an input intensity I_0 that travels in a doped waveguide, the output intensity I is expressed by [26]:

$$I = I_0 \exp[g(\nu)L] \tag{5.38}$$

where L is the length of the waveguide, and $g(\nu)$ is the gain of the waveguide determined by

$$g(\nu) = (N_2 - N_1) \frac{c^2 G(\nu)}{8\pi n^2 \nu^2 \tau} \tag{5.39}$$

where N_1 and N_2 are the number of atoms per unit volume at the lower and higher energy levels, respectively, c is the velocity of light in free space, n is the refractive index, $G(\nu)$ is the normalized emission spectrum, and τ is the fluorescence lifetime. To produce a positive gain, N_2 must be bigger than N_1; that is, there is an inversion of population. The inversion of population is achieved by a pump. For rare-earth-doped waveguides, an optical pump is used. Therefore, the basic parameters that characterize rare-earth-doped waveguides are their absorption and emission spectrum, fluorescence lifetime, and amplification ability.

5.7.1 Absorption and Emission Spectra

The absorption spectrum of a rare-earth-doped waveguide gives us the information to determine the wavelengths of the pump light that can be used. The emission spectrum determines the wavelengths of the light that can be amplified.

The absorption spectrum is measured by the transmission spectrum measurement setup described in Section 5.4. To determine precisely the absorption peak wavelengths and the absorption efficiency, a proper length of waveguide must be chosen so that no absorption saturation occurs.

The setup for the emission spectrum is similar to that for the absorption spectrum, but the light source must be a laser whose wavelength is situated in the efficient absorption region of the absorption spectrum, and at the output of the waveguide a filter that cuts the pump laser is added. Typical absorption and emission spectra of waveguides doped with neodymium and erbium are shown in Chapters 3 and 6 (see Figures 3.26 and 6.21).

5.7.2 Fluorescence Lifetime

The fluorescence lifetime is an important parameter of a rare-earth-doped waveguide. This parameter relates to the optical gain of the doped waveguide. The measured fluorescence lifetime can tell us whether the doped waveguide is suitable for amplification and laser action.

The fluorescence lifetime is defined as the time when the emitted light intensity decreases to $1/e$ of its original intensity after the pump light is switched off. To measure correctly the fluorescence lifetime, two conditions must be satisfied. First, the pump light power must be high enough that all the rare-earth atoms are excited. Second, the switching time of the pump source must be much smaller than the fluorescence lifetime.

The measurement setup of fluorescence lifetime is shown in Figure 5.18. We prefer the pump source to be a pulse laser with a pulsewidth smaller than the fluorescence lifetime. A continuous laser modulated by a chopper also can be used. At the output of the waveguide, a filter is used to cut the pump light and allow the fluorescent light to pass through. The detected signal is then amplified and sent to an oscilloscope, which shows the fluorescence decay as a function of time. The

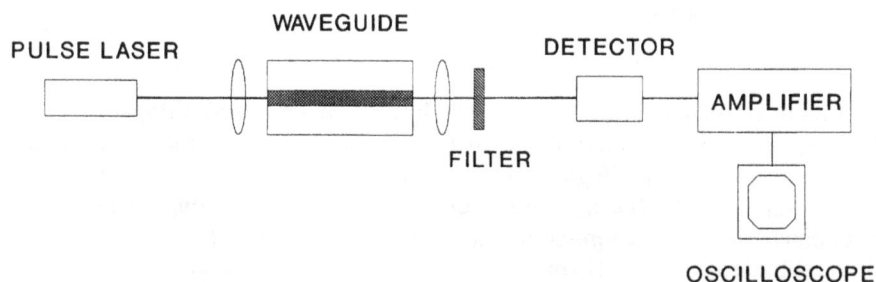

Figure 5.18 Experimental setup for fluorescence-lifetime measurement.

fluorescence lifetime is given by the time at which the emission intensity decreases to $1/e$ of its maximum value.

5.7.3 Amplification Measurement

The gain of a doped waveguide can be evaluated theoretically. It can also be determined experimentally. The experimental setup for measuring the amplification is illustrated in Figure 5.19. The pump light and the signal light to be amplified are coupled simultaneously into the doped waveguide. The signal light is modulated by a chopper. A filter that cuts the pump and lets the signal transmit is placed between the waveguide and the detector. The detected signal is sent to a lock-in amplifier. When the pump is off, we measure the signal power I_0. When the pump is on, we measure the signal power I_1. The gain is calculated as

$$g = \frac{10 \log(I_1/I_0)}{L}(\text{dB/cm}) \tag{5.40}$$

where L is the length of the waveguide.

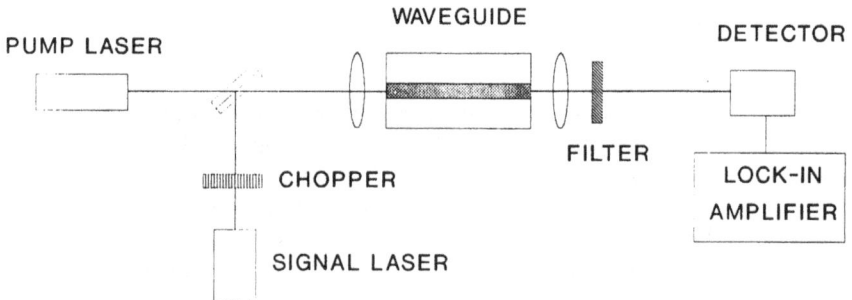

Figure 5.19 Optical amplification measurement setup.

Figure 5.20 shows an amplified signal by a 6 mm neodymium-doped waveguide [27]. The insert is the emission spectrum. The waveguide is made by silver-ion exchange in KIGRE Q-246 glass. The pump source is a dye laser operated at 580 nm. The signal is at 1.08 μm coming from a He-Ne laser. Although the 1.08 signal is not coincident with the emission peak, significant amplification is observed. The gain is estimated to be 3 dB/cm.

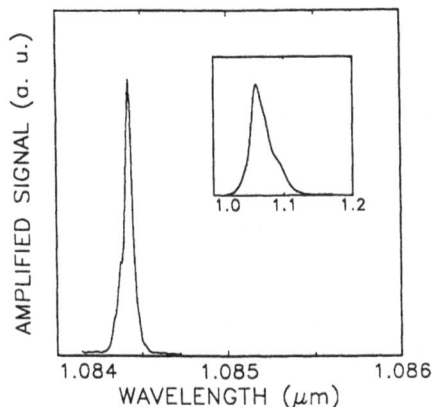

Figure 5.20 Amplified signal by a 7 mm neodymium-doped waveguide. The insert is the spontaneous emission spectrum from the waveguide.

REFERENCES

1. Tamir, T., *Integrated Optics*, Springer-Verlag, New York, 1982.
2. Ulrich, R., and R. Torge, "Measurement of Thin Film Parameters with a Prism Coupler," *Appl. Opt.*, Vol. 12, No. 12, Dec. 1973, pp. 2901–2098.
3. Seligson, J., "Prism Couplers Guided-Wave Optics: Design Considerations," *Appl. Opt.*, Vol. 26, No. 13, July 1987, pp. 2609–2617.
4. Li, M.J., S.I. Najafi, J. Albert, and K.O. Hill., "Bragg Gratings in Glass for Guided Wave Applications," SPIE Conf. 1334, paper #15, San Diego, CA, July 1990.
5. Li, M.J., S. Honkanen, W.J. Wang, R. Leonelli, J. Albert, and S.I. Najafi, "Potassium and Silver Ion-Exchanged Dual-Core Glass Waveguides with Gratings," *Appl. Phys. Lett.*, Vol. 58, No. 23, June 1991, pp. 2607–2609.
6. Gedeon, A., "Comparison Between Rigorous Theory and WKB-Analysis of Modes in Graded-Index Waveguides," *Opt. Communic.*, Vol. 12, No. 3, Nov. 1974, pp. 329–332.
7. Najafi, S.I., R. Srivastava, and R.V. Ramaswamy, "Wavelength-Dependent Propagation Characteristics of Ag$^+$-Na$^+$ Exchanged Planar Glass Waveguides," *Appl. Opt.*, Vol. 25, No. 11, June 1986, pp. 1840–1843.
8. White, J.M., and P.F. Heidrich, "Optical Waveguide Refractive Index Profiles from Measurement of Mode Indices: A Simple Analysis," *Appl. Opt.*, Vol. 15, No. 1, Jan. 1976, pp. 151–155.
9. Chiang, K.S., "Construction of Refractive-index Profiles of Planar Dielectric Waveguides from the Distribution of Effective Indices," *J. Lightwave Technol.*, Vol. LT-3, No. 2, April 1985, pp. 385–391.
10. Hertel, P., and H.P. Menzler, "Improved Inverse WKB Procedure to Reconstruct Refractive Index Profiles of Dielectric Planar Waveguides," *Appl. Phys.*, Vol. B 44, 1987, pp. 75–80.
11. Mccaughan, L., and E. Bergmann, "Index Distribution of Optical Waveguides from Their Mode Profile," *J. Lightwave Technol.*, Vol. LT-1, No. 1, March 1983, pp. 241–244.

12. Morishita, K., "Index Profiling of Three-Dimensional Optical Waveguides by the Propagation-Mode New-Field Method," *J. Lightwave Technol.*, Vol. LT-4, No. 8, August, 1986, pp. 1120–1124.

13. Coppa, G., P. Di Vita, and M. Potenza, "Two-Dimensional Index Distribution Determination from Near-Field Measurement in Single-Mode Fibre," *Electron. Lett.*, Vol. 22, No. 20, Sept. 1986, pp. 1038–1040.

14. Weber, H.P., F.A. Dunn, and W.N. Leibolt, "Loss Measurements in Thin Film Optical Waveguides," *Appl. Opt.*, Vol. 12, No. 4, April 1973, pp. 755–757.

15. Won, Y.H., P.C. Jaussaud, and G.H. Chartier, "Three-Prism Loss Measurement of Optical Waveguides," *Appl. Phys. Lett.*, Vol. 37, No. 3, August 1980, pp. 269–271.

16. Li, M.J., Doctoral Thesis, University of Nice, 1989.

17. Hunsperger, R.G., "Integrated Optics," University of Delaware, 1978.

18. Walker, R.G., "Simple and Accurate Loss Measurement Technique for Semiconductor Optical Waveguides," *Electron. Lett.*, Vol. 21, No. 13, June 1985, pp. 581–583.

19. Okamura, Y., A. Miki, and S. Yamamoto, "Observation of Wave Propagation in Integrated Optical Circuits," *Appl. Opt.*, Vol. 25, No. 19, Oct. 1986, pp. 3405–3408.

20. Haegele, K.H., and R. Ulrich, "Pyroelectric Loss Measurement in LiNbO$_3$:Ti Guides," *Opt. Lett.*, Vol. 4, No. 2, Feb. 1979, pp. 60–62.

21. Hickernell, R.K., D.R. Larson, R.J. Phelan, Jr., and L.E. Larson, "Waveguide Loss Measurement Using Photothermal Deflection," *Appl. Opt.*, Vol. 27, No. 13, July 1988, pp. 2636–2638.

22. Walker, R.G., and C.D.W. Wilkinson, "Integrated Optical Ring Resonators Made by Silver Ion-Exchange in Glass," *Appl. Opt.*, Vol. 22, No. 7, April 1983, pp. 1029–1035.

23. Boccara, A.C., D. Fournier, W. Jackson, and N.M. Amer, "Sensitive Photothermal Deflection Technique for Measuring Absorption in Optically Thin Media," *Opt. Lett.*, Vol. 5, No. 9, Sept. 1980, pp. 377–379.

24. Jackson, W.B., N.M. Amer, A.C. Boccara, and D. Fournier, "Photothermal Deflection Spectroscopy and Detection," *Appl. Opt.*, Vol. 20, No. 8, April 1981, pp. 1333–1344.

25. Jenkins, F.A., and H.E. White, *Fundamentals of Optics*, McGraw-Hill, New York, 1957.

26. Yariv, A., *Introduction to Optical Electronics*, Holt, Rinehart and Winston, New York, 1976.

27. Li, M.J., R. Leonelli, and S.I. Najafi, "Rare-Earth-Doped Glass Waveguide Optical Amplifiers," SPIE Conf. 1338, paper #11, San Diego, CA, July 1990.

Chapter 6
Waveguides and Devices
S. Iraj Najafi

Photonics Group of Montreal, École Polytechnique
Montreal, Quebec

Different ion-exchange processes for making glass waveguides were explained in Chapters 2 and 3. The optical properties of an ion-exchanged waveguide depend on the fabrication process as well as the glass substrate in which the waveguide is made. The selection of a process and a substrate is usually governed by three phenomena: mode confinement, propagation losses, and optical damage threshold.

For better mode confinement a higher maximum-index change, Δn, is required. Waveguides with large Δn can be achieved by the adequate choice of an ion-exchange process and glass substrate. Such waveguides are interesting because, for example, they can tolerate smaller bend radii. However, as we will see in this chapter, they have a smaller single-mode operation regime.

Propagation losses depend on the fabrication process as well as the quality of glass substrate. In fabricating a waveguide, special care is needed to prevent formation of metallic colloids in the waveguide region, which usually happens in fabrication of silver-ion-exchanged waveguides. One way to avoid metallic colloids was mentioned in Section 2.3.3. In Section 6.1, we will explain another way to prevent this effect, at least to some extent. The glass substrate should be of high quality with as little absorption as possible. It should have a smooth surface, free of any microcracks, to prevent scattering of propagating light.

If very high-intensity light has to propagate in a waveguide, waveguides with a low optical damage threshold cannot be used. The damage threshold, too, depends on the fabrication process. It will be discussed in more detail in Section 6.1.

Slab waveguides are the simplest waveguiding components made by ion-exchange process in glass. As explained in Chapters 2, 3, and 5, they are usually used to determine maximum index change and diffusion coefficient due to an ion-exchange process in glass. They have little practical application, and we will not discuss them in this chapter.

Channel waveguides are the building blocks of glass integrated-optical components. Depending on the simplicity (or complexity) of a device, straight, and curved as well as single- and dual-core waveguides are used. Multistep processes are also developed to make better performing devices. In this chapter we will review different approaches to making ion-exchanged glass integrated-optical components. The optical properties of the fabricated components will be discussed. We start in Section 6.1 with the simplest component, a single-core, straight-channel waveguide made by a one-step process, and proceed to more complex components.

Potassium- and silver-ion-exchanged processes have been used much more than other ion-exchange processes to make glass waveguides. We will, therefore, limit our discussion to these two processes. Depending on the device requirements, other processes (mentioned in Chapter 2) may also be employed.

6.1 STRAIGHT WAVEGUIDES

A straight waveguide is produced by an ion-exchange process through an opening in a metallic mask on the surface of the substrate (see Section 2.1). The wavelength of operation and number of modes are determined by the opening width (mask width, W) ion-exchange time, t, and temperature, T. For a single-mode waveguide fabrication in the visible and near-infrared ranges, openings of 2 μm to 10 μm are usually employed. Ion-exchange time and temperature are selected to achieve waveguide depths of about 2 μm to 10 μm as well.

Figure 6.1 depicts transmission spectra of three potassium-ion-exchanged waveguides with different mask widths (see Chapter 5 for measurement details). The waveguides are made by potassium-ion exchange in a Corning 0211 glass substrate. Ion exchange is carried out at 400°C for 5 hours. Potassium-ion exchange in this glass results in small index increase ($\Delta n = 0.006$). Consequently, the waveguides have a large single-mode operation regime. The opposite happens in silver-ion-exchanged waveguides. The value of Δn is much larger (in Corning 0211 it is 0.09), and therefore, the waveguides have a smaller single-mode operation regime. Figure 6.2 shows the transmission spectrum of a silver-ion-exchanged waveguide (Ag1) in the Corning 0211. Ion exchange is carried out through an opening of 2.5 μm in an Al mask on the surface of the substrate at 300°C for 2 hrs. Pure silver nitrate is used. Silver-ion-exchanged waveguides absorb considerable light in the visible range, due to the formation of metallic colloids under the metallic mask during the ion-exchange process [1]. This results in high propagation losses when these waveguides are operated in the visible range. However, in the infrared range, the propagation losses in silver-ion-exchanged waveguides are low, and they are

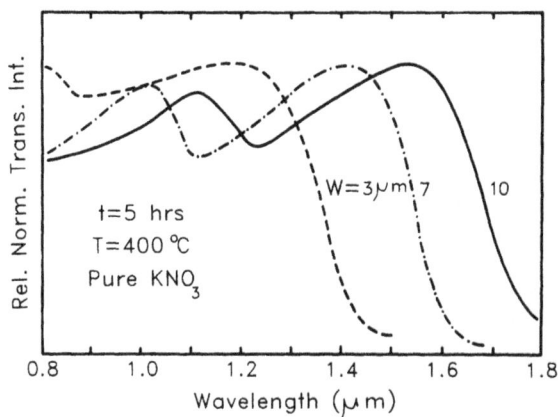

Figure 6.1 Transmission spectra of three ion-exchanged glass-channel (straight) waveguides made by using different mask openings, W.

on the same order of magnitude as those in potassium-ion-exchanged waveguides (smaller than 0.2 dB/cm at 1.3 μm) [2].

Absorption in the visible range in silver-ion-exchanged waveguides is reduced by using a two-step process. First, a potassium-ion exchange is carried out. Then, a silver-ion exchange is performed. In fact, this two-step process results in a dual-core waveguide [3]. Silver-potassium dual-core waveguides have light guiding properties superior to single-core silver-ion-exchanged waveguides. Figure 6.2 compares transmission spectrum of a silver-potassium dual-core waveguide (KAg1) with that of the silver-ion-exchanged waveguide mentioned in the previous paragraph. The same mask opening (2.5 μm) is employed to make the dual-core waveguide. The potassium-ion exchange is carried out in pure potassium nitrate at 400°C for 140 min. The silver-ion exchange is performed in pure silver nitrate at 300°C for 150 min. The dual-core waveguide essentially has the advantages of both silver and potassium single-core waveguides: low absorption in the visible range, a large single-mode regime, and good mode confinement. In addition, the silver-potassium dual-core waveguides have a higher damage threshold than single-core silver-ion-exchanged waveguides. Figure 6.3 shows the statistical fit of the probability of occurrence of damage in single- and dual-core waveguides [4]. Single-core silver-ion-exchanged waveguides break when submitted to high powers, possibly because of the residual metallic silver formed in the waveguide during the exchange process. The risk of forming metallic silver is reduced in dual-core waveguides. Single-core potassium-ion-exchanged waveguides have a much greater power-handling capability. Under the conditions used in [4], single-core potassium-ion-exchanged waveguides tolerated at least 10 times more power than their silver-ion-exchanged coun-

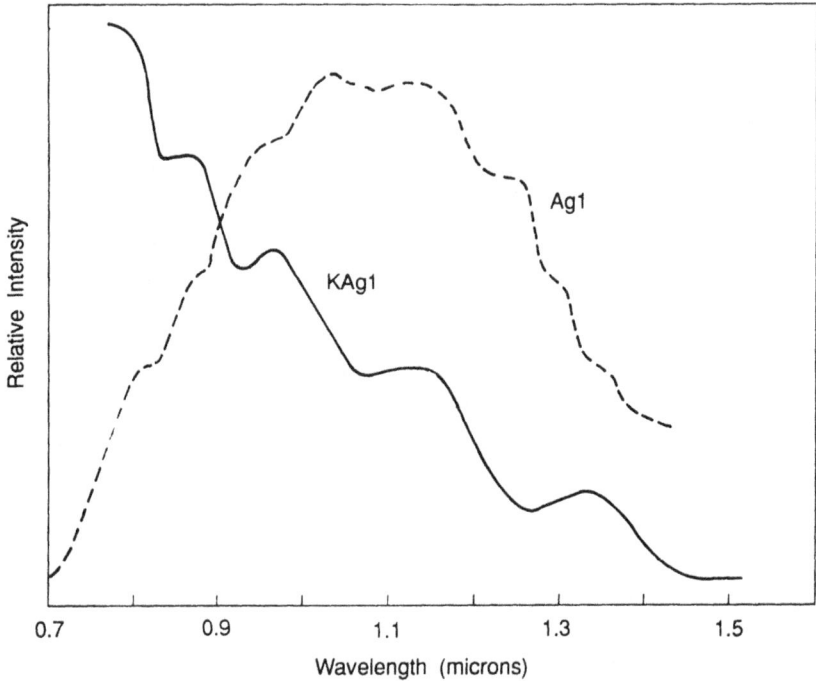

Figure 6.2 Transmission spectra of single- and dual-core waveguides. Mask width = 2.5 μm. For single-core waveguide Ag1, T = 300°C and t = 120 min. For dual-core waveguide KAg1, T = 400°C and t = 140 min. (first step); T = 300°C and t = 150 min. (second step).

terparts. In fact, it was not possible to damage the waveguides up to the maximum power available from the laser (174 mW average guided power).

Dual-core waveguides have other advantages over single-core waveguides. For example, they can have a more symmetrical transversal mode profile [3]. This is because the maximum index change in dual-core waveguides may be slightly below the glass surface due to the reduced sodium concentration near the surface after the first (potassium) ion exchange.

We may want to use the dual-core waveguides only in some parts of an integrated optical circuit. In this case, we should take the necessary measures to minimize coupling losses between single- and dual-core waveguides. A practical approach is to employ tapered waveguides [5]. Figure 6.4 depicts schematically such a device fabricated by silver- and potassium-ion-exchange processes. First, potassium exchange is used to make a Y-branch with a 6 mrad angle in pure potassium nitrate, (T = 400°C, t = 8 hrs). The resulting component is shown by

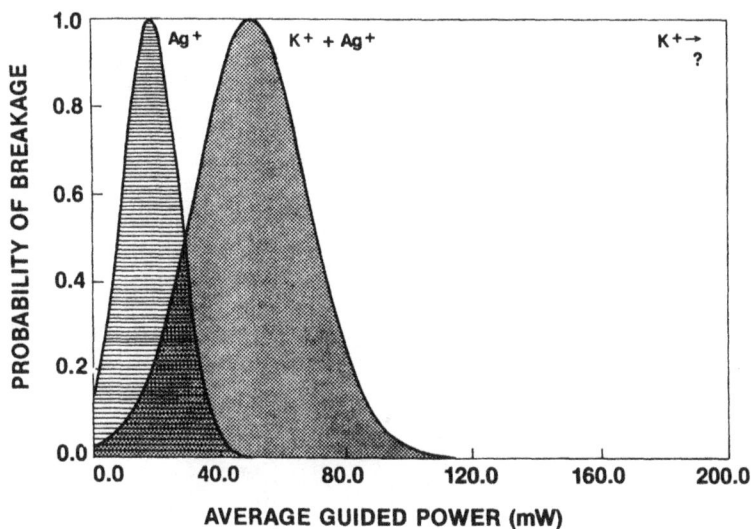

Figure 6.3 Statistical fit of the probability of occurrence of damage using a Gaussian distribution. (From [4].)

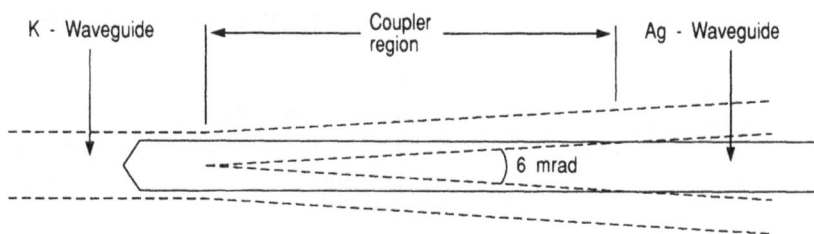

Figure 6.4 Potassium-silver waveguide coupler [5].

a dashed line. Then a silver-ion exchange is performed (silver-film technique, $T = 343°C$, voltage $= 5$ V, $t = 2$ min, annealed for 20 min), using a tapered mask. The resulting waveguide is shown by a solid line. The mask opening is 4 μm in both steps. Figure 6.5 shows the mode profiles at the 1.3 μm wavelength of the single-core potassium- and silver-ion-exchanged waveguides measured at two ends of the fabricated device. The potassium-ion-exchanged waveguide has a much larger mode profile than the silver-ion-exchanged one. A direct coupling of light from the potassium waveguide to the silver waveguide results in very high (several dB) coupling losses. By using the tapered coupler, the excess loss is determined to be

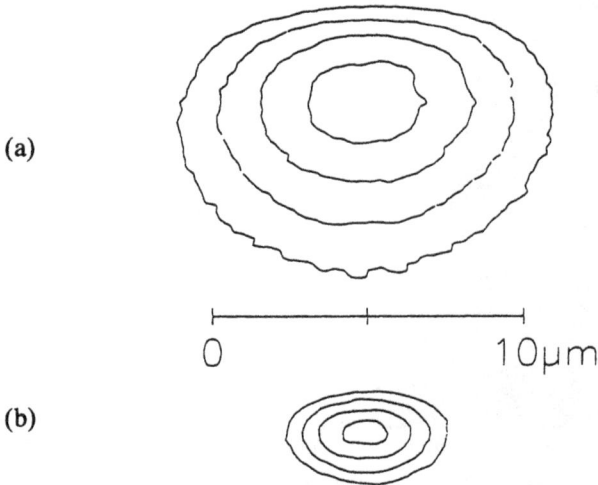

Figure 6.5 Mode intensity distribution at 1.3 μm of potassium (a) and silver (b) ion-exchanged waveguides of the coupler of Figure 6.4 [5]. Contours are for normalized intensities 0.3, 0.5, 0.7,
and 0.9.

1.0 dB. The excess loss is expected to be reduced further if we use an improved
design.

Two other methods have been employed to make waveguides with more
symmetrical profiles: burying a waveguide, and ion exchange through a barrier.
As mentioned in Section 2.3.3, to make buried waveguides, first a planar (surface)
waveguide is produced. This can be a single- or dual-core waveguide. The refractive
index of the fabricated waveguide is then modified by using one of the following
techniques:

1. The sample is placed in a molten salt that does not contain the ions used for
   waveguide fabrication. An electric field is applied across the substrate, and
   the waveguide is pulled inside the substrate (it is buried) [6]. If no electric
   field is utilized, some of the ions entered in glass during the waveguide
   fabrication can diffuse back in the melt. This reduces the concentration of
   the ions on the surface and gives rise to a buried waveguide [7]. The guided
   light is closer to the surface in this type of buried waveguide than the ones
   made by applying an electric field.

2. A simple postback can also be employed to redistribute the ions in the glass
   and move the peak of the refractive index inside the substrate. The degree
   of burial and the refractive-index profile symmetry (or asymmetry) of the
   final component depends on original mask width, temperature and time at
   each step, and the strength of the applied electric field (if any). These param-

eters have to be optimized for each fabrication process–glass substrate combination. If adequate parameters are selected, buried waveguides with rather symmetrical refractive-index profiles can be achieved [8].

Ion exchange through a barrier is an alternative technique to make buried waveguides [9]. First, a barrier is made by a first ion exchange in glass (e.g., potassium) that is quite shallow ($\sim$1 μm). Then another ion exchange (e.g., silver) is performed to fabricate the buried waveguide. Two phenomena play primary roles in this process: (1) potassium ions are practically immobile at the temperature used for silver-ion exchange, therefore, they always stay close to the surface of the substrate; (2) concentration of sodium ions is reduced on the surface during barrier fabrication, which forces silver ions to move further inside the substrate rather than accumulate on the surface. Consequently, the maximum index change takes place underneath the potassium barrier. A postback can be performed to redistribute the ions to obtain a more symmetrical refractive index profile.

The type and sequence of ion-exchange processes in a two-step ion-exchange process govern Δn, d, and the index profile of a waveguide. If the temperature during the second step is higher, the index profile achieved at the first step will be modified. Although this is needed in some cases (e.g., buried waveguides), it may not be desirable in others (e.g., dual-core waveguides). On the other hand, the diffusion coefficient (D) and Δn may be influenced by the sequence of the process. The first ion exchange modifies the glass composition and, therefore, changes the diffusion characteristics of the second process.

6.2 CURVED WAVEGUIDES

These waveguides are needed to reduce the length of integrated optical components and circuits. However, their properties can be quite different from the straight waveguides fabricated under identical conditions. The degree of difference depends on the radius of the curvature. In particular, curved waveguides are expected to have higher propagation losses and propagation constants than the straight waveguides. An experimental study has been carried out to investigate the effect of curvature radius on losses and effective indices in S-shape waveguides [10]. As shown in Figure 6.6 these components are composed of two straight waveguides connected by using a curved waveguide. The curved waveguide is designed by smoothly connecting two equal parts of circular rings having the same radii. The waveguides are made by potassium-ion exchange in Corning 0211 at 400°C using pure potassium nitrate. Ion exchange is carried out for 5 hrs. Mask width (10 μm) is maintained constant along the waveguide. Different curvature radii (20, 30, 40, and 60 mm) are employed.

Figure 6.7 depicts transmission spectra of the S-shape waveguides with different radii. A comparison of these spectra with that of a straight waveguide fabricated under identical conditions (see Figure 6.1) reveals that the curvature

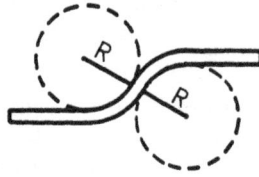

Figure 6.6 Schematics of a curved (S-shape) waveguide.

radius can have a very drastic effect on the transmission characteristics of a curved waveguide. The operating wavelengths are reduced considerably by decreasing the radii of the curvature in the waveguides. This is related to the losses due to the curvature. The propagation losses in these waveguides are measured, and the excess losses (= losses in S-shape waveguide − losses in straight waveguide) are determined. Figure 6.8 shows the results. The curvature radii of smaller than 30 mm or so introduce a lot of loss in potassium-ion-exchanged waveguides. In fact, we see that, in a two-mode S-shape waveguide with 20 mm radii, the first-order mode ($m = 1$) disappears at the bend [10]. If smaller radii are required, ion-exchange processes that result in better mode confinement (higher index change) should be used. For example, we made a silver-potassium dual-core S-shape waveguide with 20 mm radii. The waveguide had three modes at 0.6328: before, at, and after the bend. Unlike single-core potassium-ion-exchanged waveguides, all of the modes were guided through the bend.

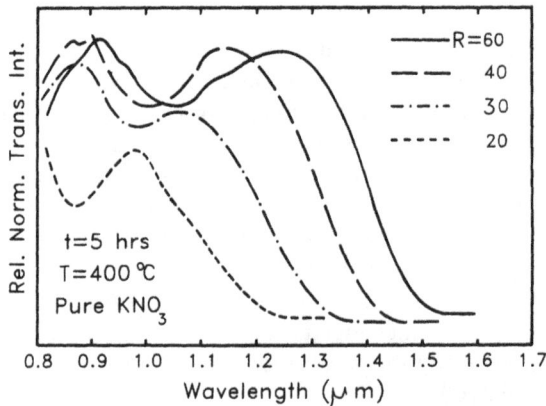

Figure 6.7 Transmission spectra of S-shape waveguides. (From [10].)

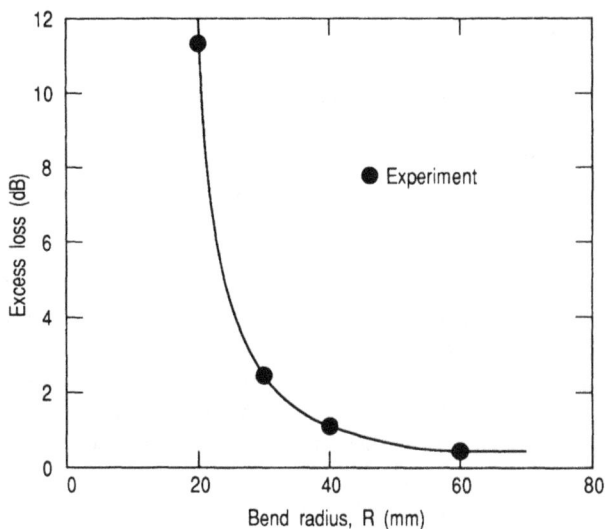

Figure 6.8 Excess loss in S-shape waveguides as a function of bend radius.

In an effort to study further the S-shape waveguides, we excited the fundamental mode of multimode (two or three modes) single- and dual-core waveguides. We used the grating coupling technique [10] to observe the number of modes before, at, and after the bend. The fundamental mode was guided through the waveguide in both structures. We did not observe any coupling of light to the other modes.

6.3 WAVEGUIDES WITH GRATINGS

Channel waveguides with gratings are useful components in the fabrication of integrated optics devices. They can be employed to couple light into or out of the waveguides for out-of-plane interconnects for instance or to filter-selected wavelengths in WDM systems. This opens up possibilities for the realization of new, efficient, and potentially inexpensive integrated optics devices.

In a waveguide with a grating, a guided mode with propagation constant β_{gm} interacts with the refractive index perturbation of period Λ and generates scattered harmonic fields with propagation constants β_h [11]. For $|\beta_h| < 2\pi/\lambda$, the diffracted fields radiate in the air and the substrate. If $2\pi/\lambda < |\beta_h| < 2\pi n_s/\lambda$, the diffracted fields radiate only in the substrate. The term n_s is the refractive index of the substrate. If $|\beta_h| > 2\pi n_s/\lambda$, the grating does not couple the guided field to the radiating modes. In the special cases where $|\beta| = |\beta_{gm}|$, the grating couples guided modes between them. In particular for $\beta_h = -\beta_{gm}$, the incident mode is reflected.

This corresponds to the usual Bragg condition $\lambda_B = 2N\Lambda$, where $N = \beta\lambda/2\pi$ is the effective index.

Three types of gratings can be made on glass waveguides: etched, deposited, and ion exchanged. The behavior of glass waveguides with etched and deposited gratings is usually similar [12]. Therefore, we will limit our discussion in this section to the waveguides with etched and ion-exchanged gratings.

Fabrication processes of etched and ion-exchanged gratings in glass waveguides are shown in Figure 6.9. A holographic setup with a HeCd laser (442 nm) is employed to register the grating in a photoresist layer. For an etched grating, a plasma etch is used to transfer the registered grating in the waveguide. For an ion-exchanged grating, a chemical etch is utilized to make a grating in an aluminum mask. A second ion exchange is carried out through the openings in this mask. A photoresist layer doped with coumarin is employed to prevent reflection of light from Al mask. Figure 6.10 shows scanning electron microscope pictures of etched and ion-exchanged gratings in glass waveguides.

Etched and silver-ion-exchanged gratings are fabricated on single- and dual-core waveguides, and their optical properties are investigated [3, 13]. Table 6.1 summarizes fabrication parameters of these components. Diffraction and transmission properties of the fabricated components are studied at 0.6328 μm and around 1.3 μm, respectively. To measure the diffraction efficiency, light from a He-Ne laser is coupled into the waveguides, and the transmitted light at the output and the light diffracted into the air by the gratings are measured. An example of

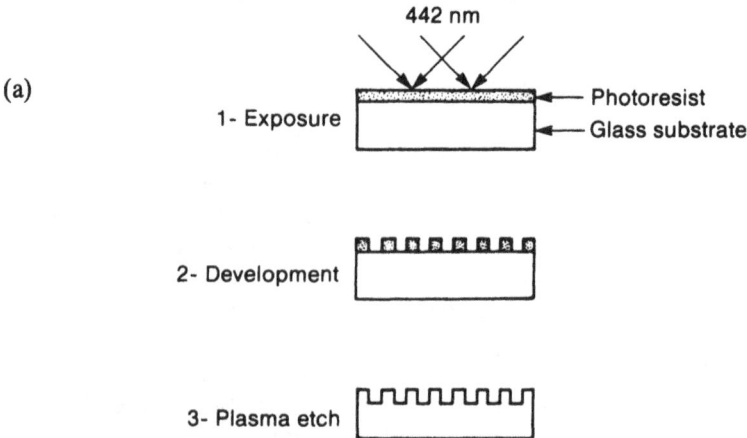

Figure 6.9 Etched (a) and ion-exchanged (b) grating fabrication processes. (From [13].)

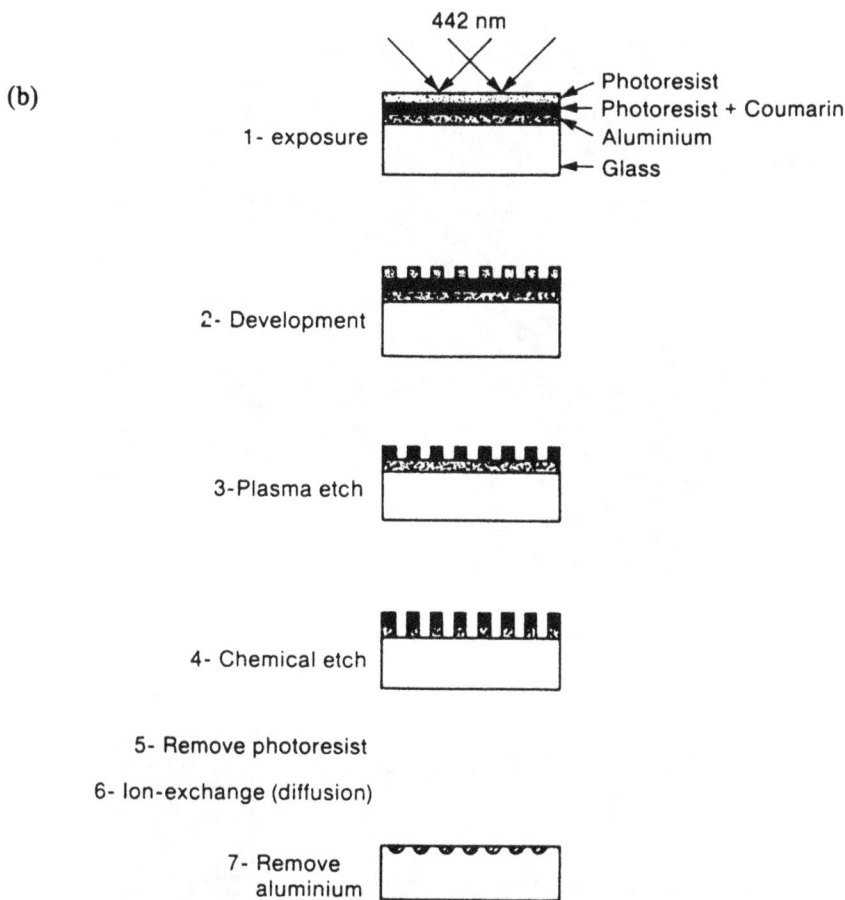

442 nm

(b)

1- exposure

Photoresist
Photoresist + Coumarin
Aluminium
Glass

2- Development

3-Plasma etch

4- Chemical etch

5- Remove photoresist

6- Ion-exchange (diffusion)

7- Remove aluminium

Figure 6.9 Continued.

the diffracted light by a grating in a waveguide is shown in Figure 6.11. The ratio of the total power in the diffracted light to that of the transmitted light is determined. The results are summarized in Table 6.1. In the sample with a shorter ion-exchange time (KAg1), the diffraction efficiency depends on the coupling at the input of the waveguide, because in this waveguide light at 0.6328 μm can be guided in silver- or potassium-ion-exchanged region.

Ion-exchanged waveguides with an ion-exchanged grating have a planar structure. Therefore, other components such as detectors can be integrated relatively easily with these components. However, their efficiency is relatively low; and they are not suitable, for example, for efficient wavelength filtering. Ion-exchanged

(a)

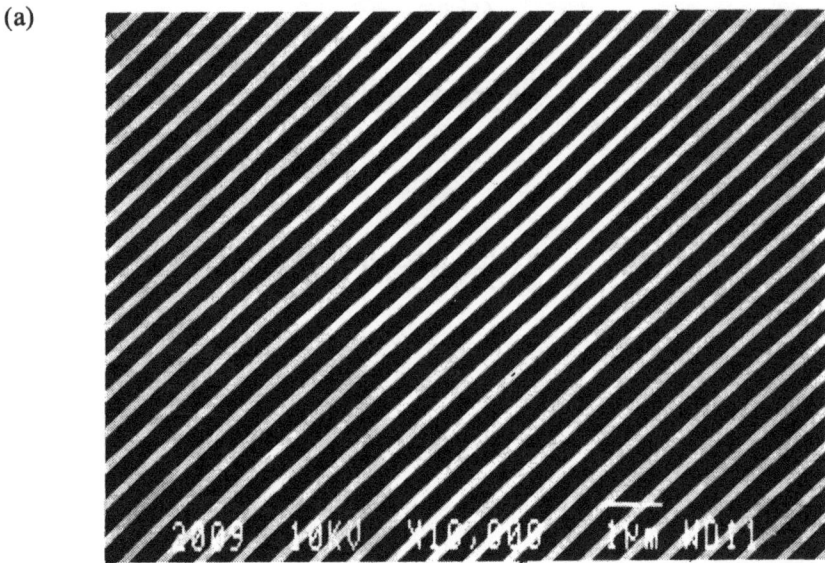

(b)

Figure 6.10 SEM pictures of etched (a) and ion-exchanged (b) gratings in glass. (From [6, 13].)

Table 6.1
Fabrication parameters of Ion-Exchanged Glass-Channel Waveguides with Grating
(Ratio of power in diffracted light in air to transmitted light is also given; $\lambda = 0.6328\ \mu$m)

| | | Waveguide | | Grating | | |
Sample	Configuration	K^+ exchange	Ag^+ exchange	Type	Ag^+ exchange	Ratio
KAg1	Dual core	400°C, 140 min.	300°C, 150 min.	Etched	—	4 to 42
KAg2	Dual core	400°C, 140 min.	300°C, 300 min.	Etched	—	83
Ag1	Single core	—	300°C, 120 min.	Etched	—	32
K1	Single core	400°C, 120 min.	—	Ion exchange	270°C, 60 min.	0.01

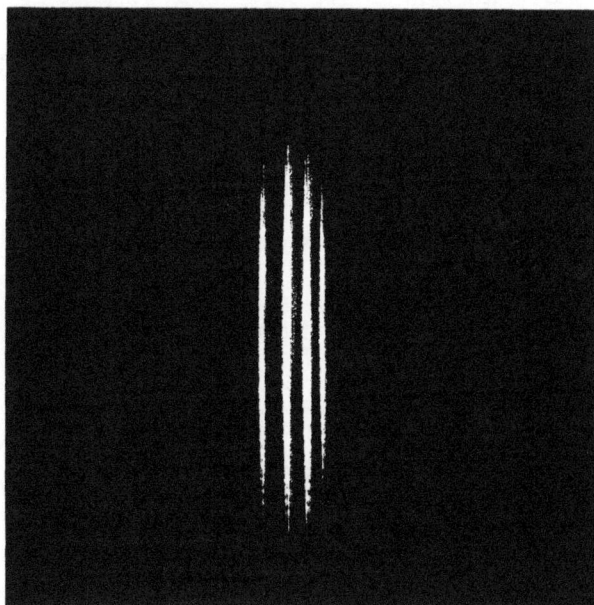

Figure 6.11 Photograph of diffracted lights by an etched grating on a single-core silver-ion-exchanged waveguide. Distance between screen and sample is 2 m.

waveguides with etched gratings exhibit a much higher efficiency. This efficiency depends on the dimensions of the grating and the waveguide. The etched gratings on all of the fabricated waveguides have a depth of about 0.15 μm. Waveguide depth in single-core potassium-ion-exchanged waveguides is 9 μm, which is about three times larger than that of silver-ion-exchanged waveguides. Consequently, the guided light has more interaction with the grating in the silver-ion-exchanged wave-guides.

To investigate the transmission properties of the waveguides with grating, light from an LED is coupled into the waveguides, and the transmitted light is measured using a spectrometer. The gratings are designed to reflect light around 1.3 μm. No significant reflection is observed in ion-exchanged waveguides with ion-exchanged gratings. In single-core potassium-ion-exchanged waveguides with etched gratings reflection is relatively weak (10% or so). In single-core silver-ion-exchanged and in dual-core waveguides more than 95% reflectivity is observed for the TE modes. Figure 6.12 depicts transmission spectra of TE and TM modes of the dual core waveguide KAg1. At the wavelength of reflection the waveguide is single mode and light is guided in the silver-ion-exchanged region. For TM polar-ization, the reflectivity is much smaller (less than 10%). This may be explained by the difference in coupling coefficients for TE and TM polarizations:

$$C_{TE} = \frac{\omega \mathcal{E}_0}{4P} \iint (n^2 - n_0^2)\, \vec{E}_t \cdot \vec{E}_t^*\, dx\, dy \tag{6.1}$$

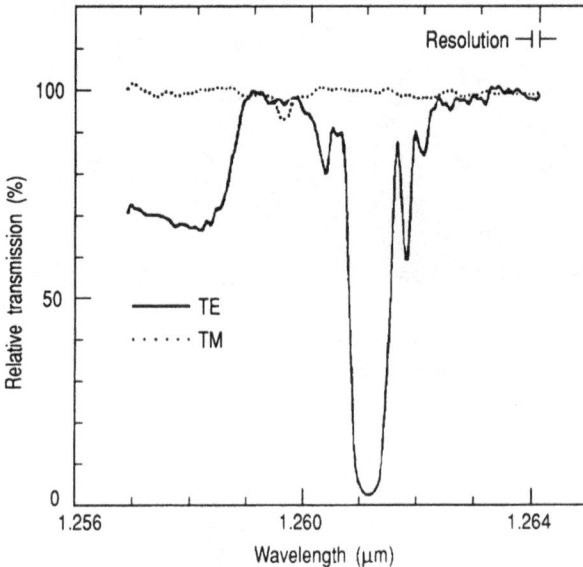

Figure 6.12 Transmission spectra of TE and TM modes of dual-core waveguide KAg1.

$$C_{\text{TM}} = \frac{\omega\mathcal{E}_0}{4P}\left[\iint (n^2 - n_0^2)\,\vec{E}_t \cdot \vec{E}_t^*\,dx\,dy - \iint \left(\frac{n_0^2}{n^2}\right)(n^2 - n_0^2)\,\vec{E}_z \cdot \vec{E}_z^*\,dx\,dy\right] \qquad (6.2)$$

where

P = power in a guided mode,
E_t = transversal field,
E_z = longitudinal field,
n_0 = refractive index (without grating),
n = refractive index in grating area.

If the second term in eq. (6.2) is large, the coupling coefficient for the TM mode is small and consequently the reflection is low. A theoretical study of TE and TM reflectivities in ion-exchanged waveguides with etched grating was recently carried out [14]. The results confirm the experimental observations. The unequal reflectivity for TE and TM polarizations in the ion-exchanged waveguides with grating renders them suitable candidates for polarization filter as well. Very compact filters with reasonably good efficiencies should be possible.

6.4 COUPLERS

Couplers are employed to combine or divide guided light in integrated optical circuits. Figure 6.13 shows three basic structures to perform these functions. Each structure has certain advantages and drawbacks. The Y-branch is probably the simplest coupler and can be used for single- or multimode operation. However, it exhibits a minimum intrinsic loss of 3 dB as a single-mode combiner and is not practical for splitting ratios different from 1:1. Directional couplers are more interesting because they can be used to divide the guided light at any ratio. Their main drawback is their high sensitivity to the device parameters and fabrication factors. To achieve low-loss Y-branches and directional couplers it is usually necessary to design a smooth transition section between the input and output waveguides or select very small separation angles. These components may then become long. In star couplers one waveguide is connected to three or more waveguides. Again, to ensure a low-loss operation and achieve an equal power division among the waveguides it is essential to make a smooth transition section. It may also be necessary to tailor the refractive index in this section. These problems complicate the fabrication procedure. For this reason, we usually prefer to cascade Y-branches or directional couplers to make 1 × n power dividers-combiners. In this section, we will discuss in some detail Y-branches and directional couplers.

(a)

(b)

(c)

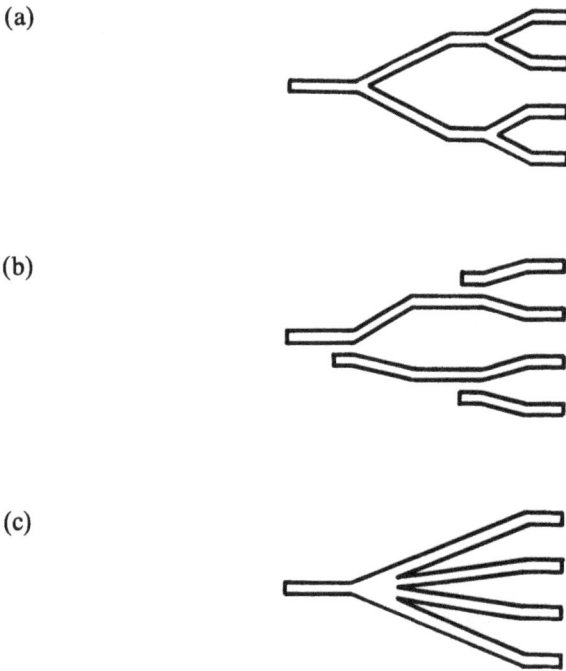

Figure 6.13 Three integrated optical couplers for power division-combination: (a) Y-branch; (b) directional coupler; (c) star coupler.

6.4.1 Y-Branch

As mentioned previously Y-branches are usually employed to divide guided light into two waveguides in an equal ratio. By cascading them, light can be split into 4, 8, 16, or more equal parts. They are also utilized in some special cases, such as Mach-Zehnder interferometers, as power combiners because the coherence of the combining light can be used to avoid the 3 dB extra loss.

The excess loss (loss due to the Y-branch) in a Y-branch power divider depends primarily on the separation angle. To achieve an excess loss of, say, 0.5 dB (or less) a separation angle of less than 1 degree or so should be used [8]. To reduce the length of a Y-branch, curved waveguides can be employed after the branching point. Then, the optical characteristics of such a Y-branch are governed mainly by (1) the transition section, which is usually multimode, and effects such as mode interference may take place; and (2) the radius of curvature in the output waveguides. Figure 6.14 depicts a transmission spectrum from one of the two output ports of a 1×2 Y-branch coupler with S-shape output waveguides. The radii for the *S*-shape waveguide are 30 mm. Other parameters are identical to those of the straight waveguide (made by using a 10 μm mask opening) discussed in Section 6.1. A comparison of the transmission spectrum of the *Y*-branch with that of the straight waveguide (see Figure 6.1) reveals significant differences in characteristics of the *Y*-branch and the straight waveguide [10]. In particular, the cutoff wavelength decreases and extra oscillations appear. The decrease in the cutoff wavelength is due to the curved waveguides (as we saw in Section 6.3). The oscillations most probably are due to the modal interference in the multimode transition section.

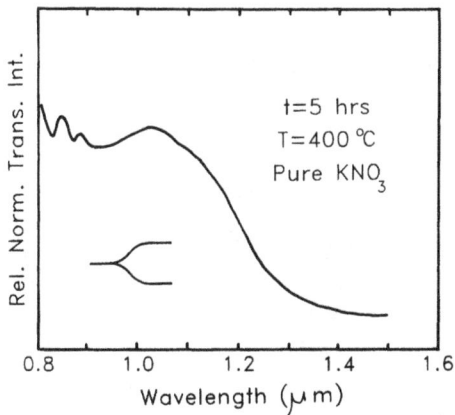

Figure 6.14 Transmission spectrum of a *Y*-branch coupler with *S*-shape output waveguides.

6.4.2 Directional Couplers

Power division and wavelength demultiplection are probably the most important applications of passive glass directional couplers. In this section we will explain two examples of such devices, examine their performance, and discuss the tolerance required in their fabrication.

Symmetrical directional couplers are employed to divide guided power. The mask used to make the couplers has the following parameters (see Figure 6.15(a)):

Mask width for each waveguide = 5 μm.

Separation between waveguides (center to center) = 11 μm.

Interaction length L_c = 0.1, 0.3, 0.5, . . . , 1.9 mm.

Separation angle = 1.0°.

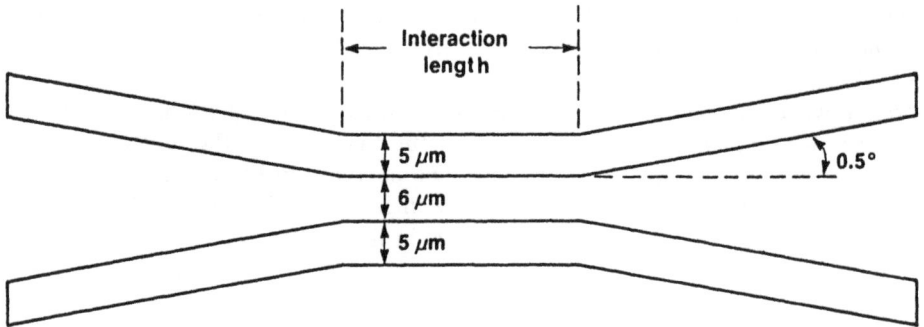

(a)

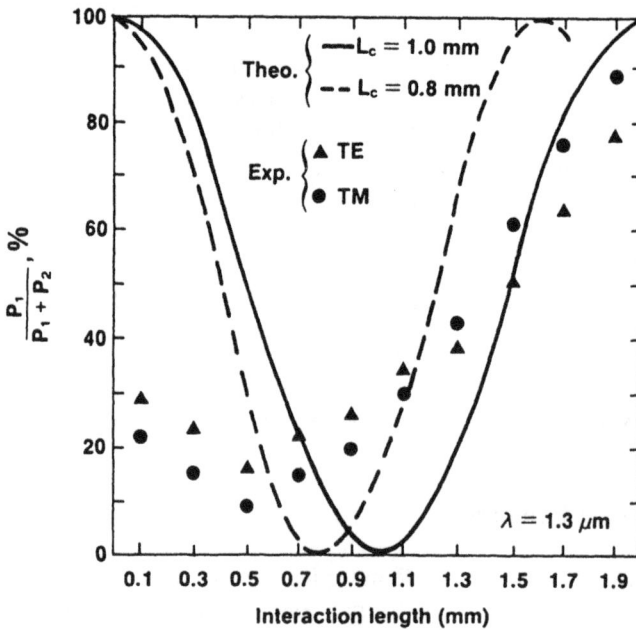

(b)

Figure 6.15 (a) Schematic of directional coupler mask. (b) Variation of power division with interaction length. Solid and dashed lines are plotted using eq. (6.3).

The couplers are fabricated by potassium-ion exchange in Corning 0211, using pure potassium nitrate. Ion exchange is carried out at 400°C for 90 minutes. The waveguides fabricated under these conditions are single mode in the 1.05 μm to 1.35 μm region.

Figure 6.15(b) shows how the power division ratio varies with the interaction length in these couplers [15, 16]. P_1 is the power remaining in the input waveguide, and P_2 is the power transferred to the other waveguide. In a symmetrical, ideal directional coupler, the power division ratio between two output parts is given by

$$\frac{P_1}{P_1 + P_2} = \cos^2\left(\frac{\pi z}{2 L_c}\right) \tag{6.3}$$

where z is the propagation direction and L_c is the coupling length (total power transfer length). This equation is plotted for different values of L_c in Figure 6.15(b). We can see significant differences between the theoretical prediction and the experimental results. There are two reasons for this discrepancy: (1) due to the small separation angle, considerable power transfer takes place before and after the separation points; (2) a phase mismatch exists between the two waveguides because of a small difference in the opening (mask) width (0.1 to 0.2 μm). A more realistic formula may be used by defining an effective coupling length $z + \Delta z$ and a phase factor $x = \delta/2K$. Then, eq. (6.2) takes the following form:

$$\frac{P_1}{P_1 + P_2} = \cos^2\left[\frac{\pi}{2 L_c} \sqrt{1 + x^2} \, (z + \Delta z)\right]$$
$$+ \frac{x^2}{1 + x^2} \sin^2\left[\frac{\pi}{2 L_c} \sqrt{1 + x^2} \, (z + \Delta z)\right] \tag{6.4}$$

The fit of this formula to the experimental results is shown in Figure 6.16.

In these couplers, power division and excess loss are measured at different wavelengths in the single-mode regime. We see that both of these values stay fairly constant (within 5%) in this region [15]. Excess loss is ~0.5 dB.

The power division ratio in nonsymmetrical couplers can be designed to be wavelength sensitive [17]. This may be employed to make a wavelength demultiplexer. Asymmetry can be achieved by using different dimensions or refractive indices for the two waveguides of a directional coupler. To make waveguides with different refractive indices one should employ, for example, different ion-exchange processes for each waveguide. This requires multistep photolithography and ion-exchange processes. Using two waveguides with different dimensions is much simpler. It can be achieved simply by using different mask openings for the waveguides.

To produce a demultiplexer with small bandwidth and high extinction ratio, it is better to employ different mask openings and ion-exchange processes to fabricate the two waveguides of the coupler. Such a device has a high tolerance and is very sensitive to fabrication factors.

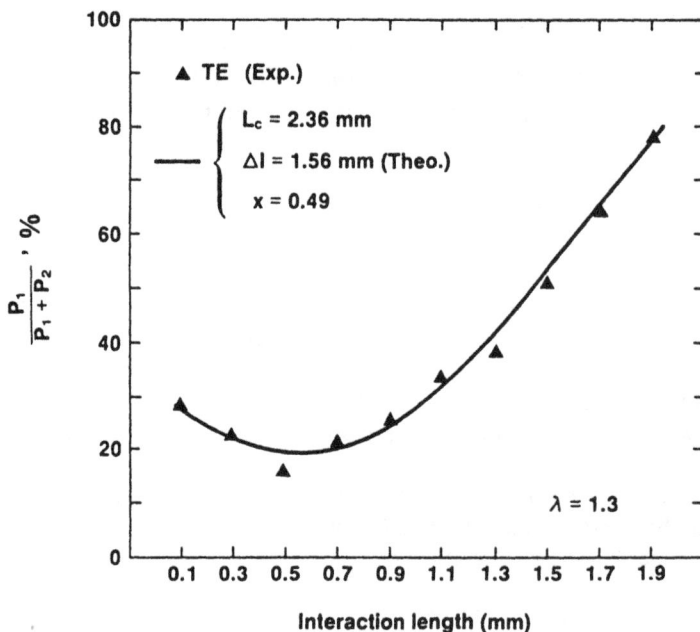

Figure 6.16 Fit of eq. (6.4) to the experimental results of Figure 6.15.

An alternative is to use a nonplanar structure to make nonsymmetrical directional couplers having two waveguides with different dimension and refractive indices. Two different ion-exchange processes (say, silver and potassium) should be employed. However, this technique requires only one photolithography process. The procedure is as follows:

1. Make a channel waveguide by the potassium-ion-exchange process.
2. Pull this waveguide inside the substrate (bury it) by using a field-assisted process.
3. Employ the silver-ion exchange to make another waveguide through the same mask opening.

With an adequate choice of fabrication parameters in the three ion-exchange processes, nonsymmetrical directional couplers with desired behavior can be achieved.

6.5 MACH-ZEHNDER INTERFEROMETERS

A schematic diagram of a Mach-Zehnder interferometer is shown in Figure 6.17. The guided light is divided into two optical waveguides (arms) and, after traveling

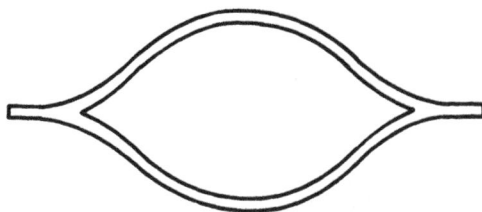

Figure 6.17 Symmetrical Mach-Zehnder interferometer.

in these two waveguides, are recombined by a coupler into a single waveguide. If the interferometer is symmetrical, there is no phase difference between the light that arrives in both couplers, and they combine constructively to produce maximum output. However, if a phase difference is introduced between the light guided into the two arms, the output is different and expressed by

$$I_{out} = \frac{I_{in}}{2} (1 + \cos \phi) \tag{6.5}$$

where the angle ϕ is the phase difference between the light guided in two arms of the interferometer. This effect can be employed to produce a number of practical devices, such as pressure and refractive index sensors and all-optical switches. In all of these devices a change in one arm is used to produce the phase difference. In a pressure sensor, the pressure is applied to one of the arms. In a refractive index sensor, one arm is exposed to a liquid. In an all-optical switch, a nonlinear material (e.g., polymer) is deposited on one of the arms. Then, the refractive index of this material is modified by increasing the intensity of the guided light. If the change in the refractive index is sufficient to produce a phase difference of π between the two arms, no light will exit the output.

The characteristics of a Mach-Zehnder interferometer is governed by the parameters of the components employed in its construction [10]: straight and S-shape waveguides and couplers. The radii of the S-shape waveguides should be large enough to prevent excessive losses. The separation angle of the coupler should not be too large again to minimize excess loss. If Y-branch couplers are utilized, a tapered section may be needed to reduce excess losses at the branching points.

More complex ion-exchanged integrated-optical Mach-Zehnder interferometers have been proposed. Figures 6.18(a–c) show schematically three examples. The first interferometer (Figure 6.18(a)) has two arms of different lengths. It has interesting applications, for example, in making temperature sensors and wavelength filters. Because the two arms have different lengths, a change in the temperature will change the phase difference, which can be measured and related to the temperature change. The phase difference between two arms depends on wavelength. The differences in the path length can be selected to produce a phase change of π at a wavelength, thus, filtering that wavelength. The bandwidth of this filter can be reduced by cascading two or more interferometers.

Figure 6.18(b) depicts a Mach-Zehnder type wavelength demultiplexer. It consists of a 3 dB coupler [18]. The light to be demultiplexed is split into the interferometer arms by a symmetric splitter. Light interferes in a four-port hybrid optical coupler, which consists of a symmetric Y-branch and an asymmetric adiabatic Y-branch. The symmetric part of this coupler combines the evenly split light

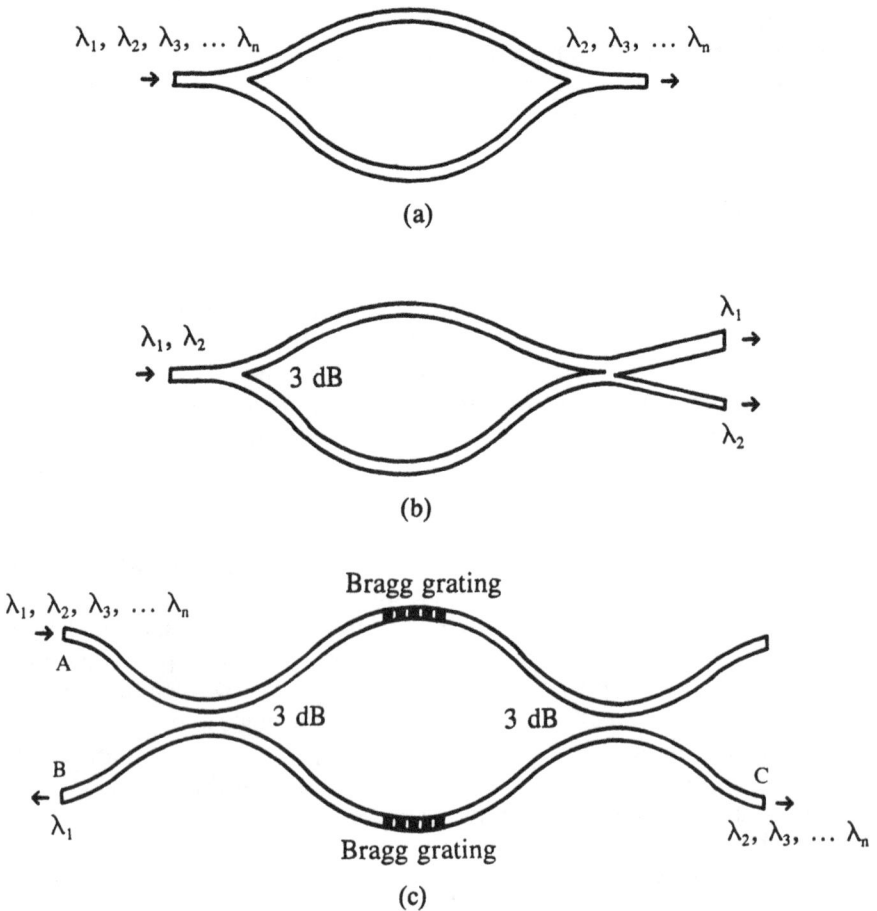

(a)

(b)

(c)

Figure 6.18 (a) Asymmetrical Mach-Zehnder interferometer, (b) and (c) wavelength demultiplexers.

into the local guided modes of the coupler junction. If the phase difference in the two interferometer arms is 0, all light is coupled to the local fundamental mode. With a phase difference of π, light is coupled to the higher-order, antisymmetric mode. With other values for the phase difference, optical power is divided to both modes according to this phase difference. The nonsymmetric branch of the coupler acts as a mode splitter: with proper adiabatic conditions the fundamental mode is coupled to the output waveguide with a higher effective index, the antisymmetric mode is coupled to the other waveguide. The device is fabricated by a silver-film ion-exchange process in Corning 0211 glass substrate. The initial branching angle is 3 mrad. Mask openings along the device are 4 μm (for input waveguide and arms), 2.5 μm and 5.5 μm (for output waveguides). The total length of the device is about 13 mm with about 3 μm path difference between two arms. Figure 6.19 shows measured output powers from the two nonsymmetric waveguides. A crosstalk of less than -20 dB is achieved in demultiplexing 1.3 μm and 1.5 μm guided lights.

A combination of gratings and Mach-Zehnder interferometers have also been proposed to make glass integrated-optical wavelength demultiplexers [19]. The device, which is shown schematically in Figure 6.18(c), has a much higher wavelength selectivity than previously discussed demultiplexer. The bandwidth for each wavelength depends on the fabrication parameters of the grating. As we saw in Section 6.3 very high reflectivities over very short bandwidths can be achieved easily. The device operates as follows: non-Bragg-resonant-wavelength light con-

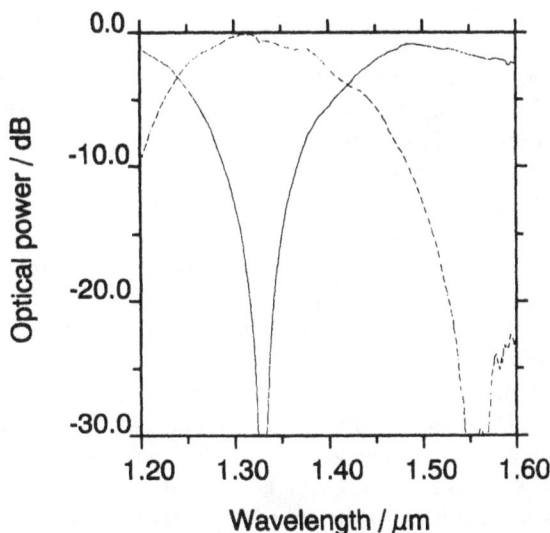

Figure 6.19 Measured output powers of the demultiplexer of Figure 6.18(b) as a function of wavelength.

taining several discrete wavelengths launched into input *A* is divided equally by the first 3 dB coupler, propagates via the two paths to the second coupler, recombines, and exits through output port *C*; Bragg-resonant light is reflected by the reflection filters, coherent recombination occurs, and the light exits via port *B*. The lights reflected back to port *A* cancel each other due to the relative phase change accumulated in passing through the coupler.

A thermooptic effect can be used to create a phase difference between the light propagating in two arms of a glass Mach-Zehnder interferometer. This is achieved by changing the refractive index of one branch with respect to the other. This technique is employed to make thermally tuned attenuators [20]. An ion-exchanged Mach-Zehnder interferometer is made in glass. A heater in aluminum is evaporated on one arm of the interferometer. Attenuation of up to 25 dB is achieved for both TE and TM polarizations.

6.6 AMPLIFIERS AND LASERS

Glass waveguides doped with rare earths can be used to make optical amplifiers and lasers. As a beam of light passes through a rare-earth-doped waveguide, energy from the incident photon promotes an ion from the ground state to a more highly excited energy state (pumping). The ion in the upper states can relax to the lower energy states either radiatively or nonradiatively. Radiative decay to the lower energy states takes one of two forms: spontaneous or stimulated emission. Spontaneous emission always takes place when the ions are in the upper excited states. If the lifetime of the excited states exceeds that of the pump states by a significant margin, population inversion takes place and stimulated emission is more probable than stimulated absorption. The light passing through the waveguide can then be amplified.

Laser oscillation requires a resonant cavity in which the space between the two mirrors contains the rare-earth-doped waveguide with a population inversion. The light initially produced by a photon from spontaneous emission reflects back and forth along axis of the laser cavity and is amplified by stimulated emission of other photons. Dielectric mirrors and Bragg grating are used to form the cavity [21].

Rare earths have absorption and emission wavelengths throughout visible and near infrared ranges [22]. It is possible to pump rare-earth-doped waveguides by laser diode and obtain amplifiers and lasers in the visible and infrared ranges. Figure 6.20 shows an energy level diagram and lasing transition of trivalent rare earths in glass [23]. Neodymium-doped waveguides have attracted more attention, probably due to their high efficiency and room temperature operation. In addition, neodymium's emission around 1.3 μm is interesting in optical communication. These characteristics result from the four-level energy structure of neodymium.

The emission of erbium around 1.5 μm is also interesting in optical communication. However, erbium has a three-level energy structure, which causes

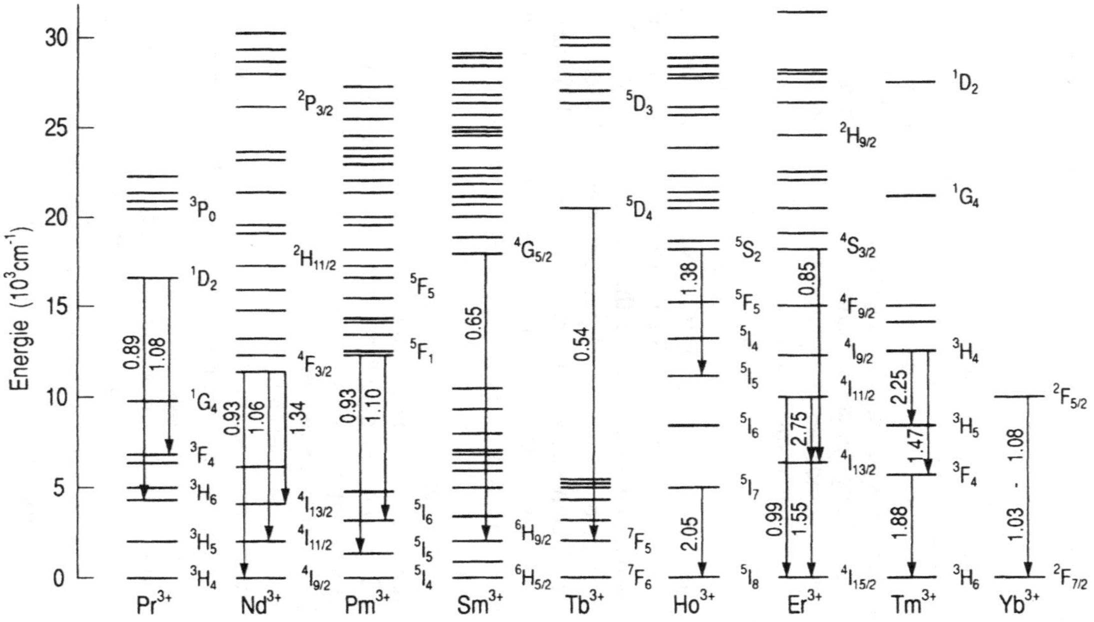

Figure 6.20 Energy level diagram and lasing transition of trivalent rare earths in glass.

competition between the absorption and the emission of lasing photons. It is expected to have a relatively higher threshold power.

Ion exchange has been the most popular technique to make rare-earth-doped waveguides in glass (to review other techniques see [24]). Most of the research activities until now have been concentrated on neodymium-doped waveguides. Silver- and potassium-ion-exchange processes in silicate and phosphate glasses have been employed to demonstrate amplifiers and lasers around 1.06 μm, 1.08 μm, and 1.3 μm [25–30]. Molten salts are used to make waveguides in silicate glasses. The silver-film technique is more convenient for waveguide fabrication in phosphate glasses. Phosphate glasses are usually damaged when they are exposed to molten salts. Very resistant glasses with high melting temperatures and special fabrication processes such as ion exchange under vacuum (to avoid moisture) are required to make ion-exchanged waveguides in phosphate glasses by molten salt technique.

Figure 6.21 shows the transmission and emission spectra of a silver-ion-exchanged waveguide in a neodymium-doped glass substrate. The absorption line around 0.81 μm is probably the most convenient one to optically pump the neodymium-doped waveguides. Laser diodes at this wavelength are readily available.

Amplification in the neodymium-doped waveguides is achieved by simply coupling the pump and signal wavelengths in the waveguides. Table 6.2 summarizes the characteristics of some of neodymium-doped amplifiers and lasers. The lasers are made by placing (or by directly making) two mirrors at the two ends of the waveguides.

It is important to note that research and development in the field of glass waveguide lasers and amplifiers is in its early stages. Until now, researchers have concentrated on demonstration of these components, and little work has focused on optimizing their performance. For example, the substrates used for neodymium-doped waveguide fabrication are not optimized for the ion-exchange process. As reported in Table 6.2, fluorescence lifetime decreases considerably as the result of ion exchange. By using adequate substrates and optimizing the fabrication process, it should be possible to produce better performing devices.

Potassium- and silver-ion-exchange processes have also been employed to make waveguides in erbium-doped glasses [31, 32]. Potassium ion is exchanged in a molten salt to make waveguides in an erbium-doped silicate glass [31]. A silver-film technique is employed to produce waveguides in an erbium-doped phosphate glass [32]. The waveguides demonstrated interesting absorption and emission properties. Further work is needed to achieve amplifiers and lasers.

6.7 OTHER DEVELOPMENTS

Some other developments have also taken place in the field of ion-exchanged glass integrated optics. For example, ring resonators are made by silver-ion exchange and reasonable finesse is achieved [2].

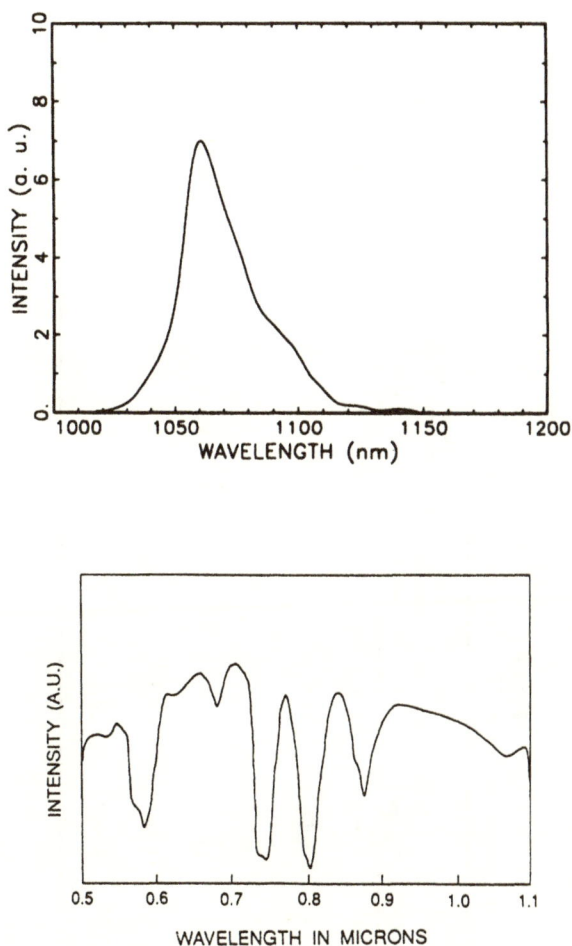

Figure 6.21 Transmission and fluorescence spectra of a neodymium-doped waveguide.

An important issue in making devices for practical application is their inter-connection to optical fiber. It appears that the general trend in the field of glass waveguides is utilization of silicon *V*-grooves. To facilitate this process and reduce the tolerances, it is helpful to taper the waveguides at the ends of the chips. A practical way to do so is by heating the end of the waveguide [33]. Higher coupling efficiencies to optical fibers were achieved by using this process. Coupling may increase further if the end of the fiber is also tapered.

Table 6.2
Characteristics of Neodymium-Doped Glass Waveguides, Amplifiers, and Lasers

Fabrication process	Ref.	Component	Nd. Conc. wt %	Length (mm)	Lifetime (μs) Substrate	Lifetime (μs) Waveguide	λ (μm) prob or lasing (pump)	Threshold or gain
Silver-ion exchange in a borosilicate glass	[25]	Laser	4	4	?	?	1.06(0.590)	18 μJ
Silver-ion exchange in Kigre Q-246 glass	[26]	Amplifier	3.5	7	300	100	1.08(0.590)	Gain of 3dB
Silver-ion exchange in Nd-doped soda lime silicate glass	[27]	Laser	?	10	?	?	1.06(0.528)	30 mW
Potassium-ion exchange in Nd-doped BK-7	[28]	Laser	2	17	?	380	1.06(0.807)	7.5 mW
Silver-film ion exchange in Hoya LHG-5	[29]	Laser	3.5	?	?	?	1.05(0.802)	6.9 mW
Silver-film ion exchange in Hoya LHG-5	[30]	Laser	3.5	5	?	?	1.325(0.802) 1.355(0.802)	33 mW

Recently, glass waveguides have been combined with polymers to produce composite waveguides [34]. This opens the possibility for making new and more sophisticated components [35]. For example, nonlinear polymers can be used to cover glass waveguides to produce all-optical devices. Couplers, interferometers, waveguides with gratings, and ring resonators are among interesting configurations for this purpose.

A lift-off technique is also employed to incorporate semiconductor detectors and lasers with ion-exchanged glass waveguides [36, 37]. Combination of glass waveguides with other materials and components provide an attractive technology for high-performance integrated optics.

REFERENCES

1. Walker, R.G., C.D.W. Wilkinson, and J.A.H. Wilkinson, "Integrated Optical Waveguiding Structures Made by Silver Ion-Exchange in Glass: 1. The Propagation Characteristics of Strip Ion-Exchanged Waveguides; A Theoretical and Experimental Investigation," *Appl. Opt.*, Vol. 22, 1983, pp. 1923–1928.

2. Viljanen, J., and M. Leppihalme, "Analysis of Loss in Ion-Exchange Glass Waveguides," Proc. First European Conf. Integrated Optics, 1981, p. 18.

3. Li, M.J., S. Honkanen, W.J. Wang, R. Leonelli, J. Albert, and S.I. Najafi, "Potassium and Silver Ion-Exchanged Glass Waveguides with Gratings," *Appl. Phys. Lett.*, Vol. 58, 1991, pp. 2607–2609.

4. Albert, J., W.J. Wang, and S.I. Najafi, "Optical Damage Threshold of Ion-Exchanged Glass Waveguides at 1.06 μm," Conf. on Integrated Optical Circuits, paper #4, 1991, Boston.

5. Honkanen, S., P. Pöyhönen, A. Tervonen, and S.I. Najafi, "A Waveguide Coupler for Potassium and Silver Ion-Exchanged Waveguides in Glass," Conf. Optical Fiber Communication (OFC'92), 1991, paper #FA1.

6. Najafi, I., and M.J. Li, "Ion-Exchanged Glass Integrated Optical Components," invited paper, Proc. First Intl. Workshop on Photonics, Components and Applications, 1990, Montebello, pp. 46–55.

7. Ramaswamy, R., and S.I. Najafi, "Planar, Buried Ion-Exchanged Glass Waveguides: Diffusion Characterization," *IEEE J. Quant. Electron.*, Vol. QE-22, 1986, pp. 883–891.

8. Okuda, E., I. Tanaka, and T. Yamasaki, "Planar Gradient-Index Glass Waveguide and Its Applications to a 4-Port Branched Circuit and Star Coupler," *Appl. Opt.*, Vol. 23, 1984, pp. 1745–1748.

9. Li, M.J., S. Konkanen, W.J. Wang, S.I. Najafi, A. Tervonen, and P. Pöyhönen, "Buried Glass Waveguides by Ion-Exchange Through Ionic Barrier," Conf. on Micro-Optics, The Hague, 1991.

10. Najafi, S.I., P. Lefebvre, J. Albert, S. Honkanen, A.V. Shahidi, and W.J. Wang, "Ion-Exchanged Mach-Zehnder Interferometers in Glass," *Appl. Opt.*, accepted for publication.

11. Tamir, T., *Integrated Optics*, Springer-Verlag, New York, 1982.

12. Lamouche, G., and S.I. Najafi, "Accurate Analysis of Ordinary and Grating Assisted Ion-Exchanged Glass Waveguides," *Opt. Eng.*, Vol. 30, 1991, pp. 1365–1371.

13. Li, M.J., and S.I. Najafi, "Fully Planar Ion-Exchanged Glass Channel Waveguides with Grating Taps," *Intl. J. Optoelectron.*, Vol. 6, 1991, pp. 575–577.

14. Li, M.J., and S.I. Najafi, "Polarization Dependence of Grating Assisted Waveguide Bragg Reflectors," *Appl. Opt.*, submitted for publication.

15. Najafi, S.I., and C. Wu, "Single-Mode Polarization-Insensitive Wavelength-Flattened Ion-Exchanged Glass Directional Couplers," Conf. Integrated Optics and Optical Communication (IOOC'89), 1989, Kobe, Japan.

16. Najafi, S.I., and C. Wu, "Potassium Ion-Exchanged Glass Waveguide Directional Couplers at 0.6328 μm and 1.3 μm," *Appl. Opt.*, Vol. 13, 1989, pp. 2459–2460.

17. Cheng, H.C., R. Ramaswamy, and R. Srivastava, "Modelling of Directional Coupler Wavelength Demultiplexers in Graded-Index Waveguides Using Normal Modes," *Tech. Dig. Integrated Photonics Research*, March 1990, p. 14.

18. Tervonen, A., P. Pöyhönen, S. Honkanen, and M. Tahkokorpi, "A Guided-Wave Mach-Zehnder Interferometer Structure for Wavelength Multiplexing," *IEEE Photon. Tech. Lett.*, Vol. 3, 1991, pp. 516–518.

19. Najafi, S.I., K.O. Hill, F. Bilodeau, and D.C. Johnson, "Method for Making a Grating-Assisted Optical Waveguide Device," U.S.A. patent #4963177.

20. Jackel, J.L., J.J. Veselka, and S.P. Loman, "Thermally Tuned Mach-Zehnder Interferometer Used as a Polarization Insensitive Attenuator," *Appl. Opt.*, Vol. 24, 1985, pp. 612–614.

21. Najafi, S.I., and K.O. Hill, "Optical Waveguide Device and Methods for Making Such Device," U.S.A. patent, 5080503.

22. Najafi, S.I., W.J. Wang, and R. Leonelli, "Ion-Exchanged Rare-Earth-Doped Waveguides," Intl. Conf. in Optical Science and Engineering: Glasses and Optoelectronics, 1989, Paris, paper #24.

23. Weber, M.J., Proc. Second Intl. Symp. New Glasses, 1989, p. 55.

24. Najafi, S.I., "Rare-Earth-Doped Waveguides for Integrated Optics," invited paper, Cost 217 Intl. Workshop on Active Fibers, Helsinki, 1991.

25. Saruwatari, N., and T. Izawa, "Nd-Glass Laser with Three Dimensional Optical Waveguide," *Appl. Phys. Lett.*, Vol. 24, 1974, pp. 603–605.

26. Li, M.J., and S.I. Najafi, "Rare-Earth-Doped Glass Waveguides and Amplifiers," *SPIE Proc.*, Vol. 1338, 1990, pp. 82–87.

27. Stanford, N.A., K.J. Malon, and D.R. Larson, "Integrated Optic Laser by Field Assisted Ion-Exchange in Neodymium Doped Soda Lime Silicate Glass," *Tech. Dig. Integrated Photonics Research*, 1990, p. 114.

28. Mwarania, E.K., "Low Threshold Monomode Ion-Exchanged Waveguide Laser in Neodymium Doped BK-7 Glass," *Electron Lett.*, Vol. 26, 1990, pp. 1217–1218.

29. Aoki, H., O. Marayama, and Y. Asahara, "Glass Waveguide Laser," *IEEE Phot. Tech. Lett.*, Vol. 2, 1990, pp. 459–460.

30. Aoki, H., O. Marayama, and Y. Asahara, "Glass Waveguide Laser Operated Around 1.3 μm," *Electron Lett.*, Vol. 26, 1990, pp. 1910–1911.

31. Jackel, J.L., A. Yi-Yan, E.M. Vogel, and A. Von Lehmen, "Guided Blue and Green Upconversion Fluorescence in an Erbium- and Ytterbium-Containing Silicate Glass," *Tech. Dig. Integrated Photonics Research*, 1991, p. 64.

32. S. Honkanen, S.I., Najafi, P. Pöyhönen, G. Orcel, W.J. Wang, and J. Chrostowski, "Silver-Film Ion-Exchanged Single-Mode Waveguides in Er-Doped Phosphate Glass," *Electron Lett.*, Vol. 27, 1991, pp. 2167–2168.

33. Mahapatra, A., and J.M. Connors, "Thermal Tapering of Ion-Exchanged Channel Guides in Glass," *Opt. Lett.*, Vol. 13, 1988, pp. 169–171.

34. Schlotter, N.E., J.L. Jackel, P.D. Townsend, and G.L. Baker, "Fabrication of Channel Waveguides in Polydiacetylenes: Composite Glass/Polymer Structures," *Appl. Phys. Lett.*, Vol. 56, 1990, pp. 13–15.

35. Najafi, S.I., and S. Honkanen, "Ion-Exchanged Glass Waveguides for Nonlinear Applications," invited paper, Electrochemical Society Meeting, Symp. on Nonlinear Optics and Materials, Toronto, 1992.

36. Chen, W.K., A. Yi-Yan, T.J. Gmitter, and J.L. Jackel, "Integration of InGaAs/InP p-i-n Photodetectors with Glass Waveguides," *Tech. Dig. Conf. Integrated Photonics Research*, 1990, p. 53.

37. Yi-Yan, A., K. Chan, T.S. Ravi, J. Gnitter, R. Bhat, and K.H. Yoo, "Grafted In GaAsP Light Emitting Diodes on Glass Channel Waveguides," *Electron Lett.*, Vol. 28, 1992, p. 341.

Index

Absorption spectrum, 132–133
Amplification measurement, 134–135
Amplifiers, 160–162

Beam propagation method, 101–104

Channel waveguides, 56–65, 83–87
  multimode, 56–60
  single-mode, 61–65
  solving mode properties in, 97–101
Charge-controlled process
  channel waveguide, 62–65
  slab waveguide, 55–56
Cleaning considerations, 18–19
Control apparatus, 18
Couplers, 151–156
  directional, 153–156
  Y-branch, 152–153
Crucibles, 17–18
Curved waveguides, 143–145
Cut-back method, 124

Diffraction-efficiency measurement, 131
Directional couplers, 153–156
Dopant ions, 14–15
  concentration of, 24–28
Dry processes, 44–48
Duration
  effect on index change, 30–32
  effect on slab waveguides, 49–50

Effective-index method, 96–97
Effective refractive-index measurement, 107–112
  grating-coupling method, 109–112
  prism-coupling method, 108–109
Electrode configurations, 44–48
Emission spectrum, 132–133

Fabry-Perot interferometer method, 124–125
Field assisted ion exchange, 12–13
  special requirements for, 19
Fluorescence lifetime, 133–134
Furnaces, 17

Gauss-Seidel method, 100
Glass
  diffusion and migration of ions in, 74–76
  integrated optics, 2–3
  ion conductivity in, 10
  for ion exchange, 14–15
  and ion source interface, 76–78
Grating-assisted waveguide characterization,
      129–132
  grating-efficiency measurement, 131–132
  grating period measurement, 129–130
  SEM analysis, 130–131

Index change, 24–28
  profiles, 28–32
  refractive, 87–88
Index-profile determination, 114–119
  inverse WKB method, 116–118
  mode near-field method, 118–119
  WKB method, 114–115
Integrated optics, 1–2
  glass, 2–3
  ion-exchanged glass, 3–4
Ion
  mobilities, 28–30
  sizes, 15–16
  source, 76–78
  stuffing, 8
Ion current density, 50–51
Ion exchange
  chemical change, 24–26
  driving mechanisms of, 10–13
  equipment, 16–19
  materials and conditions for, 13–16
    glasses and dopant ions, 14–15
    substrate composition, 15–16
    temperatures, 15
  modeling, 73–87
  multistep processes, 19–21
  physics and chemistry of, 10–16
  postprocessing of waveguides, 21–24
  process sequence, 19
  purpose and history, 7–8
  refractive index change from, 87–88
  from salt melts, 7–37
  state of the art, 9
  stresses, 26–28
  waveguide properties, 24–33
Ion stuffing, 8

Lasers, 160–162
Loss measurement, 120–129
  cut-back method, 124
  Fabry-Perot interferometer method, 124–125
  photothermal deflection method, 127–129
  prism-coupling method, 120–123
  scattered-light method, 126–127

Mach-Zehnder interferometers, 156–160
Mode near-field method, 118–119
Mode-profile characterization, 112–114
Mode propagation constants, 91–93

Network formers, 7

One-dimensional solutions, 78–82
Optical waveguides. *See* Channel waveguides;
      Slab waveguides; Waveguides